The Renin-Angiotensin System: Current Research Progress in The Pancreas

ADVANCES IN EXPERIMENTAL MEDICINE AND BIOLOGY

Recent Volumes in this Series

Volume 682 MUSCLE BIOPHYSICS: FROM MOLECULES TO CELLS
Edited by Dilson E. Rassier

Volume 683 INSECT NICOTINIC ACETYLCHOLINE RECEPTORS
Edited by Steeve Hervé Thany

Volume 684 MEMORY T CELLS
Edited by Maurizio Zanetti and Stephen P. Schoenberger

Volume 685 DISEASES OF DNA REPAIR
Edited by Shamim I. Ahmad

Volume 686 RARE DISEASES EPIDEMIOLOGY
Edited by Manuel Posada de la Paz and Stephen C. Groft

Volume 687 BCL-2 PROTEIN FAMILY: ESSENTIAL REGULATORS OF CELL DETH
Edited by Claudio Hetz

Volume 688 SPHINGOLIPIDS AS SIGNALING AND REGULATORY MOLECULES
Edited by Charles Chalfant and Maurizio Del Poeta

Volume 689 HOX GENES: STUDIES FROM THE 20TH TO THE 21ST CENTURY
Edited by Jean S. Deutsch

Volume 690 THE RENIN-ANGIOTENSIN SYSTEM: CURRENT RESEARCH PROGRESS IN THE PANCREAS
Edited by Po Sing Leung

A Continuation Order Plan is available for this series. A continuation order will bring delivery of each new volume immediately upon publication. Volumes are billed only upon actual shipment. For further information please contact the publisher.

Po Sing Leung

The Renin-Angiotensin System: Current Research Progress in The Pancreas

The RAS in the Pancreas

Po Sing Leung, PhD
School of Biomedical Sciences
Faculty of Medicine
The Chinese University of Hong Kong
Shatin
Hong Kong
China
psleung@cuhk.edu.hk

ISSN 0065-2598
ISBN 978-90-481-9059-1 e-ISBN 978-90-481-9060-7
DOI 10.1007/978-90-481-9060-7
Springer Dordrecht Heidelberg London New York

Library of Congress Control Number: 2010933093

Printed on acid-free paper

Springer is part of Springer Science+Business Media (www.springer.com)

To my wife, Wan Chun Hu and my daughter, Choy May Leung and my son Ho Yan Leung

Preface

The human pancreas consists of two organs in one: the exocrine gland made up of pancreatic acinar cells and duct cells that produce digestive enzymes and sodium bicarbonate, respectively; the endocrine gland made up of four islet cells, namely alpha-, beta-, delta- and PP-cells that produce glucagon, insulin, somatostatin and pancreatic polypeptide, respectively. While the physiological role of exocrine pancreas is to secrete digestive enzyme responsible for our normal digestion, absorption and assimilation of nutrients, the endocrine pancreas is to secrete islet peptide hormones maintaining our glucose homeostasis. The pancreatic functions are finely regulated by neurocrine, endocrine, paracrine and/or intracrine mechanisms. Thus, dysregulation of these pathways should have significant impacts on our health and disease. Nevertheless, the underlying mechanisms by which pancreatic functions are regulated remain poorly understood.

Recent basic science and clinical studies confirm myriad physiological and pathophysiological roles of the tissue renin-angiotensin systems (RAS). Of particular interest is the recent identification of a local and functional RAS in the pancreas, which influences both its exocrine and endocrine function. Its role in the pathogenesis of pancreatic diseases including diabetes and pancreatitis is increasingly recognized, as is the therapeutic potential of RAS antagonism: RAS blockade limits disease progression of type 2 diabetes mellitus and impaired glucose tolerance, and may also protect against pancreatic inflammation. To date, no single book has focused exclusively on the pancreatic RAS, and the author believes that such a publication is long overdue. This volume is the response, being an effort to draw together the current state of knowledge in the field, from the basic research to the bedside. It is our hope that the contents will excite and inform not just experts in the field, but those working in parallel disciplines and areas.

Finally, I would like to take this opportunity to express my sincere gratitude to Dr. Max Haring, Publishing Editor and Miss Marlies Vlot, Senior Assistant of Springer for their continued support and encouragement. I also express my appreciation to my graduate students, Raymond Leung, Qianni Cheng, Fanny Ma, Cynthia Lau, Ada Suen and Yuk Cheung Chan as well as Eric Wong, my technician, for their assistance. At last but not the least, I greatly appreciate the financial support for my projects provided by Research Grants Council of Hong Kong, Merck Sharp & Dohme, Novartis and Boehringer Ingelheim. Without all of the above generous support, my book could not have been published as it stands.

Hong Kong, China
December 2009

Po Sing Leung, PhD

Prologue

In 1898, Robert Tiegerstedt first described renin as a "pressor substance". Within 80 years, the role of the circulating endocrine renin-angiotensin system (RAS) in circulatory homeostasis was well described and, by 1990, angiotensin converting enzyme inhibition was transforming the pharmacological management of cardiac failure and hypertension. It was at much the same time, however, that the existence of local tissue and cellular RAS was beginning to be described. These were exciting times: before long, we knew that local RAS played a role in cardiac growth, tissue fibrosis and wound healing, cellular inflammatory responses, transduction of load in skeletal muscle, cerebral function, insulin sensitivity, adipocyte growth and differentiation.... The list went on and on. Some of these findings were pursued: they were revealed while the various RAS antagonists (ACE inhibitors and AT1 receptor antagonists) still had sufficient remaining patent life to offer commercial incentive to the manufacturing companies.

Just as such interest was waning, however, Po Sing Leung and others began to describe a pancreatic RAS—and one with extraordinary possible implications to disease pathogenesis and management. Their findings had implications for the prevention and treatment of diabetes, and also of other "orphan diseases" fore which no other specific therapies existed (such as pancreatitis). It is a shame that such findings appeared as commercial interest was declining. Nonetheless, these findings remain important- and new life may be breathed into their pursuit by the recent launch of direct renin inhibition.

This volume offers a welcome summary of the "state of the art" of this field. I, for one, earnestly hope that it encourages some to take an active interest. There is much to be done and much potential benefit yet for the prevention and treatment of pancreatic endocrine and exocrine disease. Neither would it surprise if local RAS were found to play a powerful role in endocrine regulation elsewhere in the body. Such, in general, is the way of nature...

London, UK
December 2009

Hugh Montgomery, MD

Prologue

Just when we thought we knew everything about the renin-angiotensin system (RAS), a whole new phase of discovery has emerged. Through his work and writings, Po Sing Leung points the way for many of us to become more engaged and open-minded, particularly in pushing conceptual development in this field. This book provides rich insights into how the RAS regulates both development and regulation of endocrine and exocrine functions of the pancreas. By doing so, it helps unravel the intricacies of process and mechanism of a complex system that impacts health and disease. These paradigms are not likely to be unique to the pancreas, but also extend to other organ systems. For this reason, this important book will be of interest to many disciplines and will serve as a foundation and catalyst for future discovery in this field.

Chicago, IL
December 2009

Eugene B. Chang, MD

Contents

Part I
The Pancreas

Chapter 1
Overview of the Pancreas

1.1 Structure of the Pancreas

Structurally speaking, the human pancreas consists of two organs in one: an exocrine gland and an endocrine gland. The exocrine gland is made up of pancreatic acinar cells and duct cells that produce digestive enzymes and sodium bicarbonate, respectively. The primary function of the exocrine pancreas is to secrete the digestive enzymes responsible for normal digestion and absorption of daily foodstuffs, and finally assimilation of nutrients into our body. The endocrine gland, meanwhile, is made up of five types of secretory islet cells and secretes peptide hormones for the maintenance of glucose homeostasis. The pancreatic secretory functions are finely regulated by neurocrine, endocrine, and paracrine as well as intracrine mechanisms. In view of this fact, inappropriate activation or inactivation of the pathways mediating the pancreas's fine regulatory mechanisms has considerable impacts on health and disease. Nevertheless, the underlying local mechanisms by which pancreatic function and dysfunction are regulated remain poorly understood.

The human pancreas is a retroperitoneal organ of the upper abdomen that, on average, weighs in the range of 100–150 g and measures 15–25 cm in length. It secretes about a kilogram of pancreatic juice daily into the duodenum via the ampulla of Vater where the main pancreatic duct coalesces with the common bile duct. Anatomically, the pancreas is connected with other abdominal organs including the spleen, stomach, duodenum and colon. The pancreas is structurally divided into three parts, termed the head, body and tail. The head region of the pancreas is relatively flat and situated within the first loop of the duodenum. The tail region is in close vicinity to the hilum of the spleen; it is the only part of the pancreas that contains pancreatic polypeptide (PP) cells that produce the peptide hormone PP (see Section 2.4, Chapter 2). The body region of the pancreas has a shape that resembles a prism.

The blood supply of the pancreas depends upon several major arteries, namely the inferior pancreaticoduodenal artery, the superior pancreaticoduodenal artery, and the splenic artery. Pancreatic removal of metabolites and hormone release are mediated via the pancreaticoduodenal vein and the pancreatic vein, respectively. The pancreas is highly innervated by the pancreatic plexus, celiac ganglia, and vagus nerve. For additional information of the anatomy, function and disease of the

P.S. Leung, *The Renin-Angiotensin System: Current Research Progress in The Pancreas*, Advances in Experimental Medicine and Biology 690,
DOI 10.1007/978-90-481-9060-7_1,

pancreas, please refer to an updated and integrated textbook of basic and clinical aspect of the pancreas (Beger, 2008).

The endocrine and exocrine roles of the pancreas, though distinct, have closely interrelated physiological functions. The pancreas can be considered as four structurally distinct components: the exocrine pancreas, consisting principally of acinar cells and duct cells; the endocrine pancreas, the site of the islet cells; the blood vessels; and the extracellular space. The exocrine portion, whose major function is to produce digestive enzyme and sodium bicarbonate secretion, accounts for most, if not all, of the cellular mass of the pancreas (i.e. approximately 80% by volume). The exocrine pancreas architecture is characterized by a blind-ended ductal system, structurally analogous to a massive bunch of grapes, such that each acinus equivalent to a grape is bounded by adjacent acinar cells that secrete pancreatic juice enzymes into a blind-ended tubule or called stem. The acini are grouped into lobules with a branched network of tubules. Each acinus is composed of highly orientated, pyramidal-shaped acinar cells, with their apical membranes lining a central lumen known as an intercellular canaliculus and their basolateral membranes forming the acinar periphery. The acinar cell secretions flow into the intercellular canaliculi, which consist of ductal epithelial cells, also known as duct cells, which secrete bicarbonate, mainly in the form of sodium bicarbonate. The tiny intercalated ducts that drain the acini converge into larger intralobular ducts, which then converge into major extralobular duct. The extralobular ducts finally converge into a main collecting duct, which joins with the common bile duct before the pancreatic juice enters the duodenum (see review by Leung & Ip, 2006). For the purpose of this chapter, a highly schematic illustration of the ductal and acinar cells of the exocrine pancreas is summarized and presented in Fig. 1.1.

On the other hand, the endocrine pancreas is composed of clusters of cells formerly known as the islets of Langerhans, or more simply termed pancreatic islets, to which the secretion of a number of pancreatic peptide hormones for glucose homeostasis is attributed. The endocrine pancreas is highly vascularized and morphologically distinct from its exocrine counterpart. The islets are structurally arranged in spherical shape, and account for 1–2% by volume of the total pancreatic mass. There are five major cells which constitute the islet, namely the alpha cells (α-cells), beta cells (β-cells), delta cells (δ-cells), PP cells (also known as F-cells) and, to some extent, the epsilon cells (ε-cells), which are responsible for producing glucagon, insulin, somatostatin, pancreatic polypeptide, and ghrelin, respectively (see review by Kulkarni, 2004). The secretory functions of the islet cells are modulated by signalling factors from inside and outside the pancreas. These cells can communicate with one another and influence one another's secretion. Islet cell communication modalities include, but are not limited to, humoral communication, cell–cell communication, and neural communications. In addition, there is intimate contact between the endocrine and exocrine pancreas by means of an insulin-acinar axis (Williams & Goldfine, 1985). The physiological function of the endocrine cells and their regulatory mechanisms will be discussed further in Chapter 2. In summary, a schematic illustration of the cell types and structure of the insulino-acinar portal system of the endocrine pancreas are illustrated in Fig. 1.2.

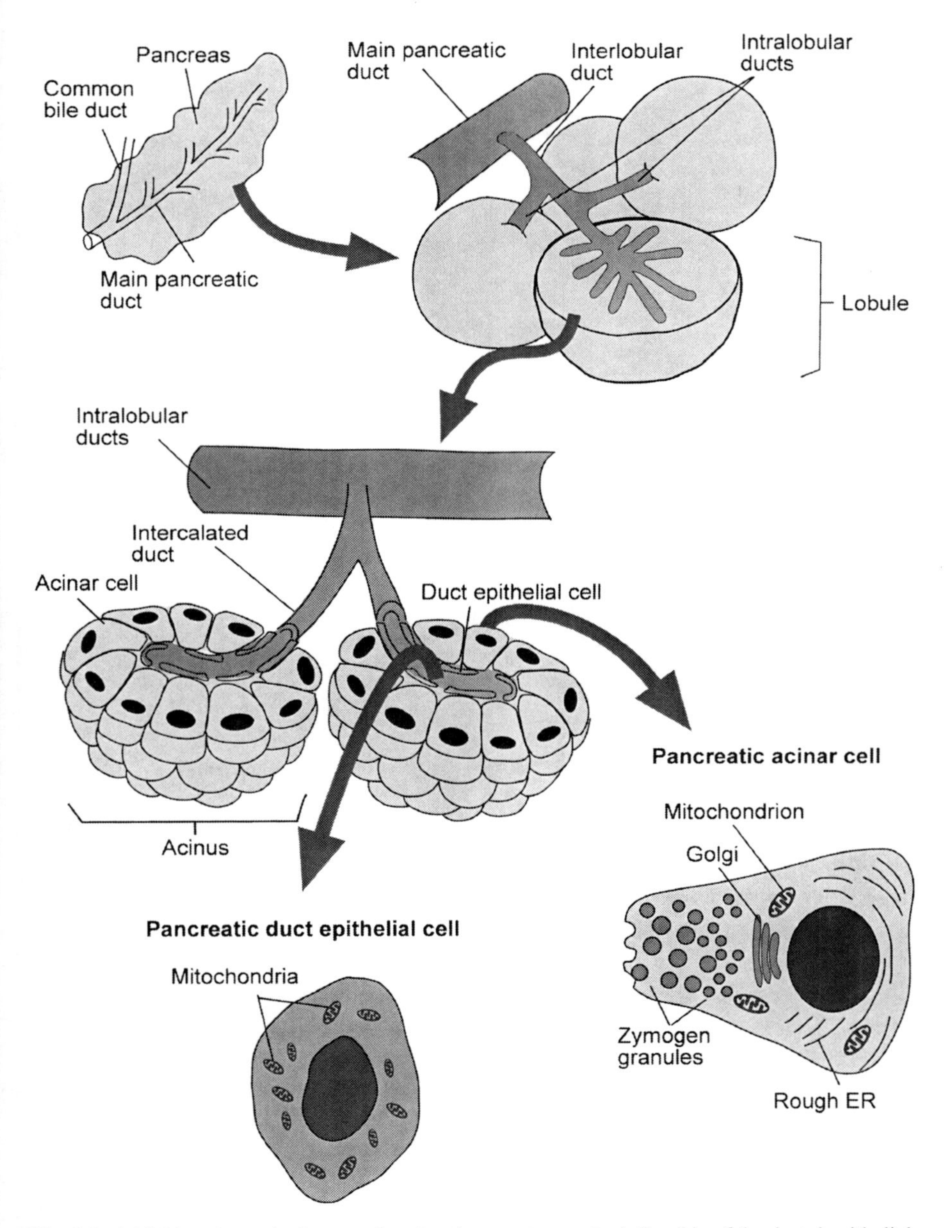

Fig. 1.1 A highly schematic diagram showing the structure and relationship of the ductal epithelial and acinar cells of the exocrine pancreas

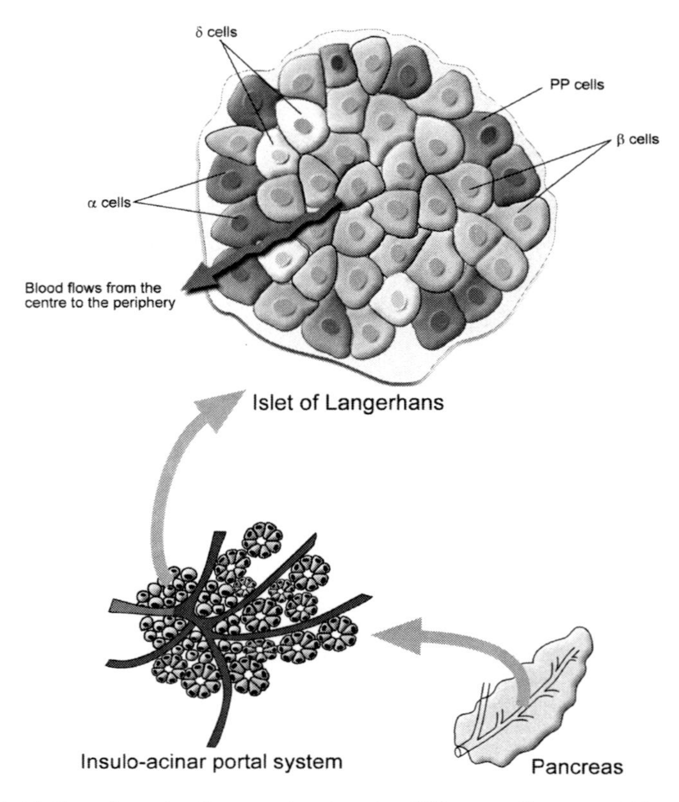

Fig. 1.2 A highly schematic diagram showing the different cell types within the islet of Langerhans and the structure of insulo-acinar portal system of the endocrine pancreas

1.2 Development of the Pancreas

Our understanding of human pancreatic development is based largely on murine model studies as the patterns of its developmental stages are well conserved between humans and mice. The developmental processes are generalized into stages of organ specification, expansion, differentiation, and maintenance, though these events overlap in time (Murtaugh, 2008). On the whole, pancreas development has been described as an epithelial-mesenchymal process wherein signals from the mesenchymal tissues govern organogenesis.

Embryologically, the development of the pancreas begins with two endoderm-lined primordial buds, or outpouchings, of the duodenum during the fourth week of human gestation. The dorsal bud extends into the dorsal mesentery to eventually form the head, body and tail of the pancreas. The ventral bud arises immediately adjacent to the hepatic diverticulum and is associated with the biliary system. Hence, the pancreas originates from two primordial buds of the primitive duodenum and is first apparent at 4 weeks gestation in humans and around embryonic day E9.5 in mice. The developmental programme of the two buds is slightly asynchronous because they receive distinct signals from the surrounding tissue context. The dorsal pancreatic bud develops in close proximity to the notochord, while the ventral bud develops in close association with the liver and under the control of signals from the overlying cardiogenic mesenchyme. The dorsal pancreatic bud grows more rapidly than the ventral pancreas, and extends into the dorsal mesentery by 6 weeks gestation. As the stomach and the duodenum rotate, the ventral and dorsal pancreatic buds fuse with the ventral bud, giving rise to the posterior part of the head and the dorsal bud, thereby forming the remainder of the organ. This fusion occurs during the seventh week of gestation in humans and around E12–E13 in mice (Suen, 2007).

The dual origin of the organ is responsible for regional differences in the distribution of islet cells in the adult: the dorsal pancreas is rich in glucagon-containing islets while the ventral portion contains PP-rich islets and few glucagon-producing cells. Fusion of the ventral duct with the dorsal duct results in the formation of the main pancreatic duct, termed the duct of Wirsung, which runs through the entire pancreas. The proximal end of the dorsal pancreatic duct usually does not communicate with the main duct, but rather forms the accessory pancreatic duct (Bonner-Weir & Smith, 1994). The pancreas at later stages undergoes a process of rapid branching morphogenesis. The mechanisms of these highly dynamic processes remained poorly understood until a recent study using real-time imaging reported visualization of the budding and branching process (Puri & Hebrok, 2007).

In mouse, two temporal waves of endocrine differentiation, termed the first and second transitions, have been observed. The first transition occurs between E9.5 and E12.5, and involves a change in shape of the pancreatic domain and the generation of early endocrine cells with co-expression of several pancreatic hormones. The second transition occurs between E12.5 and birth and encompasses a massive endocrine differentiation and major amplifications of hormone-expressing cell numbers (Pictet et al., 1972). The mature exocrine marker amylase is first detected at E13.5. The human equivalents of these transitions have not been described, though a marked transition of the exocrine genes has been observed at 11 weeks gestation (Sarkar et al., 2008). Groups of endocrine cells develop from multipotential stem cells in the ductal epithelium at 9 weeks gestation. The pancreatic acini and the first zymogen granules appear at 12 weeks gestation. Discrete islets of Langerhans can be identified at 12 weeks, and 1 week later, large primitive islet structures expressing all four pancreatic hormones are formed. While the mechanisms of specification and the reciprocal relationships of the four types of endocrine cell (alpha, beta, delta and PP cells) within the human endocrine pancreas are relatively well described during development, ghrelin-secreting epsilon cells (ε-cells) remain poorly understood; in this regard, scattered ghrelin-positve ε-cells have been observed during this period; they start to aggregate and localize around the developing islets in subsequent gestational weeks (Andralojc et al., 2009). These data have shown that ghrelin-secreting ε-cells have an ontogenetic and morphogenetic pattern that is distinct from that of apha and beta cells during development. Developing blood vessels start to penetrate into the primitive islets at 14 weeks gestation. The islets continue to increase in size through the second and third trimester (Piper et al., 2004). Most of the islet cells develop within the tail of the pancreas and in the dorsal pancreas.

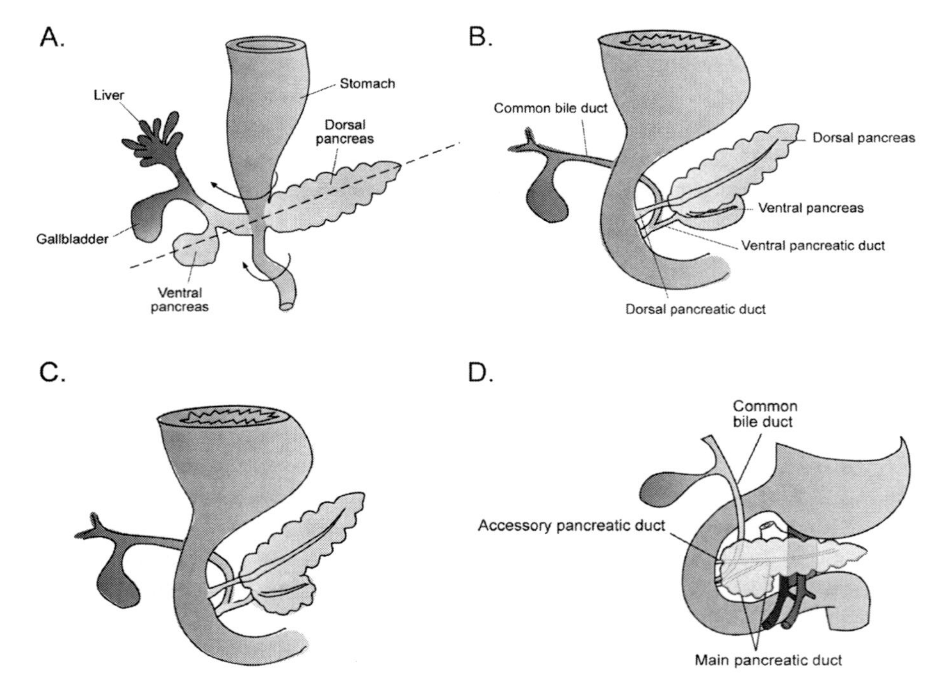

Fig. 1.3 A schematic illustration showing the different stages of progressive pancreatic development along with its corresponding gut and hepatic systems (modified from Beger, 2008)

The first cells to produce granules are the α-cells, followed soon thereafter by the β- and δ-cells. Complete maturation of the pancreatic gland does not occur until sometime after birth. Lineage tracing experiments have provided direct evidence that exocrine, endocrine islet, and duct progenitors are committed at mid-gestation and are from different lineages (Gu et al., 2002). The islet cells appear to arise from stem cells that become visible in pancreatic ducts during the third month of gestation. Islet cells migrate away from the ducts in which they arose, and move into the interlobular connective tissue. Mature islet morphology is established before birth (Kulkarni, 2004). Intriguingly, it has been more recently proposed that islet formation occurs by a process of fission following contiguous endocrine cell proliferation, rather than by local aggregation or fusion of isolated beta-cells and islets (Miller et al., 2009). Taken together, the various development stages from embryonic ventral and dorsal buds to complete formation of the pancreas and its ductal system are summarized in Fig. 1.3.

1.3 Molecular Hierarchy of the Pancreatic Development

Involvement of Wnt/β-catenin signalling in initiating pancreas development has been widely described. Inhibition of the Wnt/β-catenin pathway appears to be essential for the initial pancreatic specification during expansion of the pancreatic epithelium; such signalling mechanisms become necessary later in development (Murtaugh, 2008; Mclin et al., 2007). Pancreas generation from the gut endoderm is controlled by signals generated in the adjacent notochord. These include, among others, transforming growth factor-beta (TGFβ), activin β, and basic fibroblast growth factor (bFGF, also known as FGF2). Deficiencies in expression of these growth factors have been shown to disrupt pancreatic morphology, with epithelial structures becoming interspersed in the surrounding stroma tissue (Dichmann et al., 2003). These molecules act via repression of the expression of sonic hedgehog (Shh) in the dorsal region of the gut, the region that can subsequently form the pancreatic epithelium in the presence of pancreatic duodenal homeobox 1 (PDX-1) (Hebrok et al., 2000). Of note, the hedgehog signalling often regulates tissue morphogenesis in a dose-dependent manner. Inhibition of Shh signalling is essential in the initiation of pancreatic development in rodents. Ectopic expression of Shh can even result in transformation of pancreatic mesenchyme into duodenal mesoderm (Kawahira et al., 2003). Yet in other species, like zebrafish, Shh appears to be indispensible for proper pancreatic development (Dilorio et al., 2002).

The molecular hierarchy of pancreatic development is temporally and spatially controlled by an array of transcription factors. Their regulation determines the initial budding of the organ as well as cell fates towards the endocrine or exocrine lineage. PDX-1 (also known as IPF-1, STF-1, IDX-1 and IUF-1) plays a crucial role in the growth and differentiation of the pancreatic buds (Kaneto et al., 2008). Its expression is maintained in pancreatic precursors and becomes restricted to insulin-producing β-cells as the pancreas matures. Mice lacking this gene fail to form a pancreas. The pancreatic buds in these embryos and the dorsal bud undergo limited proliferation and outgrowth. These observations indicate that although PDX-1 defines a distinct compartment of the endodermal foregut that specifies the dorsal and ventral pancreas, additional genes must be involved in the early stages of pancreatic specification and bud formation. HNF3β (also known as Foxa2) may be involved in this early stage, as it has been shown to regulate PDX-1 promoter activity (Ben-Shushan et al., 2001). Other genes expressed early in pancreatic development, including the homeobox gene hlxb9 (encoding Hb9) and LIM-domain protein Isl1, are involved in the dorsal mesenchymal but not the ventral pancreatic programme. Following formation of the pancreatic buds and concomitant with the growth of the branching epithelial network, a second phase of transcription factor expression occurs. These factors include the basic helix-loop-helix (bHLH) proteins Ptf1/p48 (p48) and ngn3—factors that appear to control commitment towards an endocrine or exocrine lineage. Ptf1/p48 is the DNA-binding subunit of the hetero-oligomeric transcription factor Ptf1, which controls expression of genes in the exocrine pancreas (Krapp et al., 1998). Ptf1 is also required in the early patterning of undifferentiated epithelium; it is later restricted to pancreatic acinar cells. A recent study reported that an enhancer located downstream of the Ptf1 gene established an autoregulatory loop with the gene to reinforce and maintain its expression in mature acinar cells (Masui et al., 2008).

Neurogenin 3 (Ngn3) is a bHLH protein that is transiently expressed in the developing pancreas and is critical to the specification of all four types of islet endocrine cells. It is not present in mature islets. A recent study has revealed biphasic Ngn3 expression in mouse embryonic pancreas. The two waves of expression correlated exactly with the first transition and in the precedence of the second transition of endocrine differentiation (Villasenor et al., 2008). There is strong evidence indicating that Ngn3 defines an islet progenitor cell and that Ngn3-positive cells do not develop into duct cells (Gu et al., 2002). Ngn3 was demonstrated to be a determination factor for the four endocrine cells lineage in a study in which mice with disrupted Ngn3 genes all failed to generate any pancreatic endocrine cells and died postnatally from diabetes (Gradwohl et al., 2000). The expression of Ngn3 is regulated by Delta/Notch signalling on adjacent epithelial cells. Notch

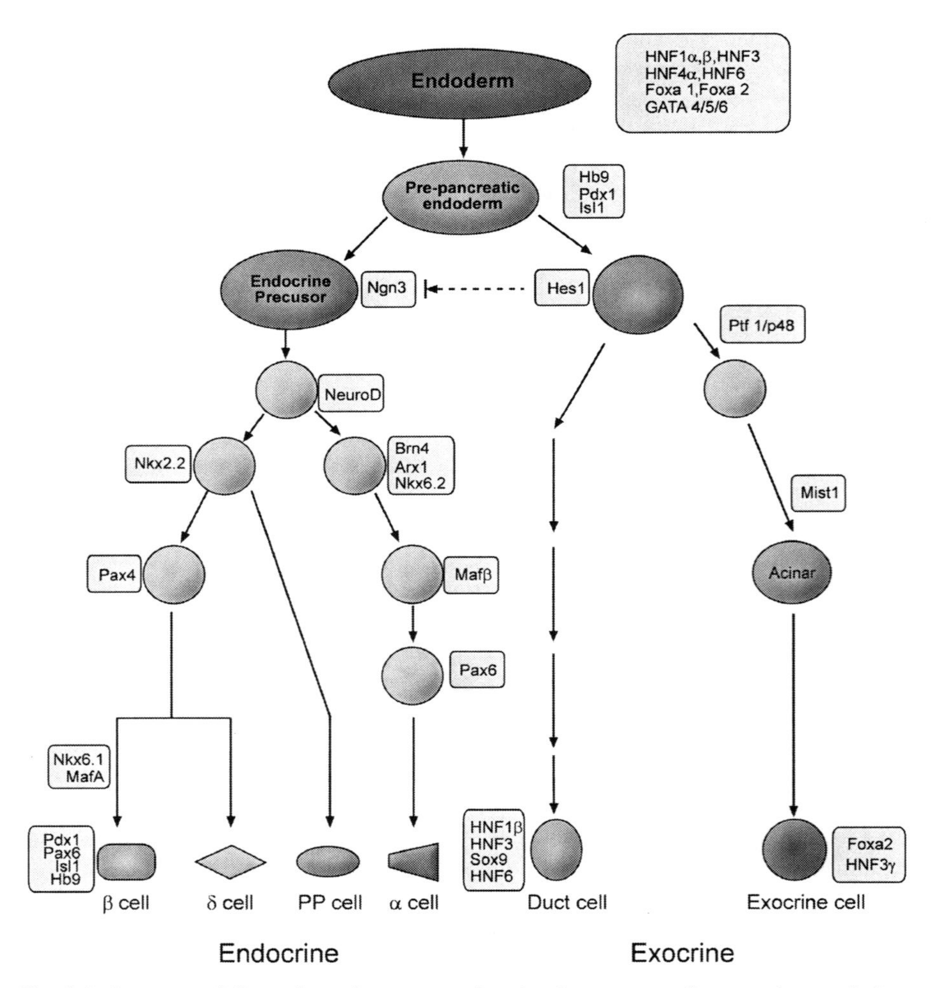

Fig. 1.4 An easy-to-follow schematic representation showing an array of proposed transcription factors that are critical in controlling development of the exocrine and endocrine cell lineage in the pancreas (redrawn from Samson & Chan, 2006)

is a transmembrane receptor and Delta is a transmembrane ligand, the expression of which is stimulated by Ngn3. When activated by Delta expressed by a close proximal cell, Notch triggers expression of another bHLH protein, Hes-1, that inhibits expression of Ngn3. Cells subjected to this inhibition of Ngn3 expression are driven towards an exocrine fate. Downstream of Ngn3 transcription factors lie a number of late-expressed transcription factors: transient expression of Nkx2.2, Pax4, Nkx6.1 and PDX-1 drives Ngn3-positive cells towards a β-cell phenotype; meanwhile expression of Brn4, Nkx6.2 and Pax6 direct cells towards an α-cell phenotype (Gradwohl et al., 2000). Any alteration in the expression of these regulatory genes will disrupt the balance of pancreatic hormone-expressing cells developed. The homeodomain protein Nkx2.2, in particular, is a critical regulator of islet specification. It has been elegantly shown that the endocrine components in Nkx2.2 mutant mice are replaced predominantly by ghrelin-expressing ε-cells (Prado et al., 2004). α-cells and a small portion of β-cells could be rescued in a Nkx2.2-Engrailed-repressor derivation (Doyle et al., 2007). For additional information about the development of the pancreas, please refer to a recent textbook (Scharfmann & Shield, 2007). Figure 1.4 summarizes a raft of proposed transcription factors that play important roles in governing the differentiation and development of the pancreatic exocrine and endocrine cells in the pancreas.

References

Andralojc KM, Mercalli A, Nowak KW, Albarello L, Calcagno R, Luzi L, Bonifacio E, Doglioni C and Piemonti L. Ghrelin-producing epsilon cells in the developing and adult human pancreas. *Diabetologia* **52**:486–493, 2009.

Ben-Shushan E, Marshak S, Shoshkes M, Cerasi E and Melloul D. A pancreatic β-cell-specific enhancer in the human Pdx-1 gene is regulated by HNF-3β, HNF-1α and SPs transcription factors. *J Biol Chem* **276**:17533–17540, 2001.

Beger HG. The Pancreas: an integrated textbook of basic science, medicine and surgery, 2nd edn. Blackwell, Oxford, 2008.

Bonner-Weir S and Smith FE. Islets of Langerhans: morphology and its implications. In CR Rahn and GC Weir (eds), Joslin's diabetes mellitus, Philadelphia, PA, pp 15–28, 1994.

Dichmann DS, Miller CP, Jensen J, Scott Heller R and Serup P. Expression and misexpression of members of the FGF and TGFbeta families of growth factors in the developing mouse pancreas. *Dev Dyn* **226**:663–674, 2003.

Dilorio PJ, Moss JB, Sbrogna JL, Karlstrom RO and Moss LG. Sonic hedgehog is required early in pancreatic islet development. *Dev Biol* **244**:75–84, 2002.

Doyle MJ, Loomis ZL and Sussel L. Nkx2.2-repressor activity is sufficient to specify alpha-cells and a small number of beta-cells in the pancreatic islet. *Development* **134**:515–523, 2007.

Gradwohl G., Dierich A, LeMeur M and Guillemot F. Neurogenin3 is required for the development of the four endocrine cell lineages of the pancreas. *Dev Biol* **97**:1607–1611, 2000.

Gu G, Dubauskaite J and Melton DA. Direct evidence for the pancreatic lineage: NGN3+ cells are islet progenitors and are distinct from duct progenitors. *Development* **129**:2447-57, 2002.

Hebrok M, Kim SK, St Jacques B, McMahon AP and Melton DA. Regulation of pancreas development by hedgehog signaling. *Development* **127**:4905–4913, 2000.

Kaneto H, Matsuoka TA, Miyatsuka T, Kawamori D, Katakami N, Yamasaki Y and Matsuhisa M. PDX-1 functions as a master factor in the pancreas. *Front Biosci* **13**:6406–6420, 2008.

Kawahira H, Ma NH, Tzanakakis ES, McMahon AP, Chuang PT and Hebrok M. Combined activities of hedgehog signaling inhibitors regulate pancreas development. *Development* **130**:4871–4879, 2003.
Krapp A, Knofler M, Ledermann B, Burki K, Berney C, Zoerkler N, Hagenbuchle O and Wellauer PK. The bHLH protein PTF1-p48 is essential for the formation of the exocrine and the correct spatial organization of the endocrine pancreas. *Genes Dev* **12**:3752–3763, 1998.
Kulkarni RN. The islet β-cell. *Int J Biochem Cell Biol* **36**:365–371, 2004.
Leung PS and Ip SP. Pancreatic acinar cell: its role in acute pancreatitis. *Int J Biochem Cell Biol* **38**:1024–1030, 2006.
Masui T, Swift GH, Hale MA, Meredith DM, Johnson JE and Macdonald RJ. Transcriptional autoregulation controls pancreatic Ptf1a expression during development and adulthood. *Mol Cell Biol* **28**:5458–5468, 2008.
McLin VA, Rankin SA and Zorn AM. Repression of Wnt/beta-catenin signaling in the anterior endoderm is essential for liver and pancreas development. *Development* **134**:2207–2217, 2007.
Miller K, Kim A, Kilimnik G, Jo J, Moka U Periwal V and Hara M. Islet formation during the neonatal development in mice. *PLoS One* **4**:e7739, 2009.
Murtaugh LC. The what, where, when and how of Wnt/beta-catenin signaling in pancreas development. *Organogenesis* **4**:81–86, 2008.
Pictet RL, Clark WR, Williams RH and Rutter WJ. An ultrastructural analysis of the developing embryonic pancreas. *Dev Biol* **29**:436–467, 1972.
Piper K, Brickwood S, Turnpenny LW, Cameron IT, Ball SG, Wilson DI and Hanley NA. Beta cell differentiation during early human pancreas development. *J Endocrinol* **181**:11–23, 2004.
Prado CL, Pugh-Bernard AE, Elghazi L, Sosa-Pineda B and Sussel L. Ghrelin cells replace insulin-producing beta cells in two mouse models of pancreas development. *Proc Natl Acad Sci U S A* **101**:2924–2929, 2004.
Puri S and Hebrok M. Dynamics of embryonic pancreas development using real-time imaging. *Dev Biol* **306**:82–93, 2007.
Samson SL and Chan L. Gene therapy for diabetes: reinventing the islet. *Trends Endocrinol Metab* **17**:92–100, 2006.
Sarkar SA, Kobberup S, Wong R, Lopez AD, Quayum N, Still T, Kutchma A, Jensen JN, Gianani R, Beattie GM, Jensen J, Hayek A and Hutton JC. Global gene expression profiling and histochemical analysis of the developing human fetal pancreas. *Diabetologia* **51**:285–297, 2008.
Scharfmann R and Shield JPH. Development of the pancreas and neonatal diabetes. Karger, Basel, 2007.
Suen PM. Isolation, characterization and differentiation of pancreatic progenitor cells from human fetal pancreas. Ph.D. Thesis, The Chinese University of Hong Kong, 2007.
Villasenor A, Chong DC and Cleaver O. Biphasic Ngn3 expression in the developing pancreas. *Dev Dyn* **237**:3270–3279, 2008.
Williams JA and Goldfine ID. The insulin-pancreatic acinar axis. *Diabetes* **34**:980–986, 1985.

Chapter 2
Physiology of the Pancreas

2.1 Exocrine–Endocrine Axis

The pancreas plays a central role in digestion and absorption as well as utilization and storage of energy substrates. As described in Chapter 1, it consists of two structurally distinct but functionally integrated glandular systems, namely the exocrine and endocrine pancreas, both of which arise from an outgrowth of the primitive gut. Secretion by the exocrine pancreas is modulated by neural and hormonal signals, particularly in the form of numerous gastrointestinal peptide hormones (Chey & Chang, 2001). Due to the lack of basal membranes or compartmentalization capsules for different cell types in the pancreas, the islets cells are interspersed within the exocrine acini. Acini located near islets, called peri-insular acini, are composed of larger sized cells possessing larger nuclei and more abundant zymogen granules than acini removed from islets, called tele-insular acini. Some secretory products of the islet cells, such as insulin, interact directly with acinar cells and thereby regulate acinar function (Murakami et al., 1992). The exclusive morphology of the peri-insular acini is reflected in the presence of high insulin concentrations in the region (von Schönfeld et al., 1994).

It has been reported that diabetes reduces pancreatic exocrine function (Chey et al., 1963), thus suggesting a close interaction between the endocrine and exocrine components of the pancreas. A novel concept of an insulin-acinar axis was proposed in 1985; it was suggested to be involved in the islet-acinar portal system and to participate in physiological regulation of acinar cell function by islet peptides (Williams & Goldfine, 1985). Pancreatic intralobular arteries branch into the islets via the vas afferents, which in turn divide into capillary glomerulus structures within the islets. The efferent vessels protrude into the surrounding exocrine pancreas to form the insulo-acinar portal system. The peri-insular acinar is hence particularly exposed to high concentrations of islet hormones in close proximity. All the efferent islet blood flows into the acinar capillary before leaving the pancreas; in doing so, no blood from the intralobular islets drains directly into veins without passing through the exocrine portion of the pancreas. Other peptide hormones produced by the endocrine pancreas, such as somatostatin, glucagon, pancreatic polypeptide (PP) and ghrelin, contribute to the regulation of pancreatic enzyme synthesis, transport and secretion, as well as to the growth of acinar cells. To better help understand this

P.S. Leung, *The Renin-Angiotensin System: Current Research Progress in The Pancreas*, Advances in Experimental Medicine and Biology 690,
DOI 10.1007/978-90-481-9060-7_2, © Springer Science+Business Media B.V. 2010

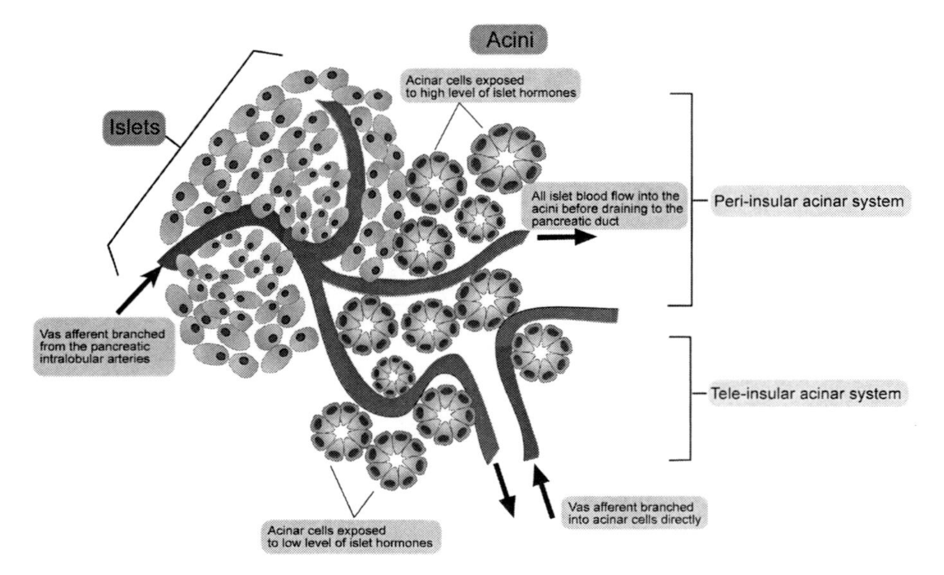

Fig. 2.1 A schematic presentation of the insulo-acinar portal system. In this system, islet and acinar cells are arranged in such an apposition that they are exposed to close contact via islet peptide hormones thus rendering exocrine and endocrine interaction to influence acinar cell function

notion, an illustration of exocrine–endocrine interactions based on the insulo-acinar portal system is shown in Fig. 2.1.

The interaction of insulo-acinar axis can be exemplified by its effects on the exocrine–endocrine functions. In this regard, the regulation of pancreatic exocrine secretion depends on the peptide hormone insulin released by the pancreatic islet β-cells. It is known that infusion of glucose into perfused rat pancreas has been shown to induce release of endogenous insulin, and thereby enhance pancreatic exocrine secretion in response to the peptide hormone CCK or cholecystokinin (Saito et al., 1980). A significant increase in pancreatic secretion was also observed in the presence of exogenous insulin (Patel et al., 2004). Interestingly, pretreatment with the muscarinic cholinergic receptor antagonist atropine abolished insulin-induced increases in pancreatic secretion in hyperglycemic rats (Patel et al., 2004). This observation implies that pancreatic secretion is stimulated by vagal cholinergic activation, which is evoked by hypoglycemia. Exogenous insulin has also been reported to inhibit pancreatic exocrine bicarbonate secretion stimulated by secretin in dogs (Berry & Fink, 1996; Howard-McNatt et al., 2002). However, hyperinsulinemia alone did not inhibit bicarbonate secretion. In contrast, hyperinsulinemia with normoglycemia reduces the bicarbonate secretion (Simon et al., 2002), demonstrating that the inhibitory effect of hyperglycemia on pancreatic secretion may be independent of insulin.

Several studies have shown that endogenous somatostatin secreted by the pancreatic δ-cells inhibits secretin and CCK release thus pancreatic exocrine secretion (Chey & Chang, 2001). Since somatostatin is present in both the intestinal mucosa and the pancreas, it is hypothesized that mucosal somatostatin may play a role in the regulation of gut hormones. Meanwhile, pancreatic somatostatin regulates pancreatic exocrine secretion directly, in the form of paracrine messengers (Beger et al., 2008). Somatostatin and PP also have suppressive roles in the insulo-acinar axis (Nakagawa et al., 1993). In contrast, the effect of glucagon on the exocrine secretion remains rather controversial. Intravenous injection of glucagon transiently can activate then suppress secretin and CCK-stimulated somatostatin, followed by an increase in circulating somatostatin in dogs (Horiuchi et al., 1993), suggesting that glucagon-mediated regulation of exocrine secretion may occur indirectly.

Functional interactions have been found between the exocrine and endocrine pancreas. For example, many diabetic patients experience changes in pancreatic exocrine function (Chey et al., 1963). However, the impaired exocrine function in these patients may not be directly linked to insulin deficiency since abnormal pancreatic exocrine functions have also been observed in some patients with type 2 diabetes mellitus (Hardt et al., 2000). Insulin deficiency in diabetic patients may be due to the inadequate actions of insulin on the potentiation of secretin and CCK-stimulated secretion, induction of amylase gene transcription (Korc et al., 1981), and down-regulation of its own receptors (Okabayashi et al., 1989).

2.2 Pancreatic Acinar Cells

The pancreatic acinar cell belongs to the exocrine portion of the pancreas. The pyramid-shaped, serous acinar cell is the dominant cell type in the pancreas, constituting 82% of the total volume of the pancreas (Bolender, 1974). Its major function is to produce, store and secrete three major categories of digestive enzymes: α-amylase, lipase and protease, which are responsible for the digestion of carbohydrates, fats and proteins, respectively. Acinar cells within each acinus are linked by gap junctions. The extensive network of ductal system of the exocrine pancreas (see Fig. 1.1, Chapter 1) allows both chemical and electrical intercellular communications to occur among acinar cells (Iwatsuki & Petersen, 1978).

An acinar cell consists of a basally located nucleus, abundant mitochondria, free ribosomes and rough endoplasmic reticulum, and contains a well developed Golgi apparatus. This organelle architecture enables production of abundant secretory granules, called zymogen granules (ZGs), which are composed of a multitude of enzymes and precursor enzymes that are polarized at the apical side of the acinar cells. ZG protease precursors include trypsinogen, chymotrypsinogen and procarboxypeptidases, which are secreted along with the isotonic NaCl mixture called pancreatic juice. Upon its release into the duodenum, inactive trypsinogen is converted into active trypsin by an enzyme called enterokinase, which is anchored to the apical membrane of enterocyte. The duodenally bound enterokinase can specifically cleave 6 amino acids from the N-terminal of trypsinogen to produce a 223-amino acid protein containing trypsin. Positive feedback by trypsin can auto-catalytically further cleave trypsinogen into trypsin and activate other precursor proteases into their respective active forms, such as chymotrypsinogen into chymotrypsin, procarboxypeptidase into carboxypeptidase and proelastase into elastase. Physiologically speaking, activation of proteases in the duodenum rather than within the pancreas prevents auto-digestion of the pancreas, thus leading to pancreatitis. In addition to proteases, the acinar secretion also contains active digestive enzymes such as pancreatic lipase, amylase and nuclease, as well as other secretory products such as co-lipase, the latter being not a hydrolytic enzyme itself (Beger, 2008). Acinar secretory products need to be directed to the intercellular canaliculi, and the ductal system, and finally into the gut lumen. Ca^{2+}-activated Cl^- channels are found specifically on the apical plasma membrane of acinar cells. Opening of these channels releases Cl^- ions into the acinar lumen; the increased negative charge leads to the influx of extracelluar Na^+ through tight junctions between acinar cells. NaCl in the acinar lumen then osmotically draws in water and thus constitutes the acinar fluid (Petersen, 2007).

There are some basic mechanisms that govern the regulation of pancreatic exocrine secretory functions. While the primary stimulus is the food itself, neural and hormonal systems play a central role in regulating secretion of acinar cells via neurocrine, endocrine and paracrine signalling pathways. The major regulators include the neurocrine molecule acetylcholine (ACh) and the endocrine molecule cholecystokinin (CCK), as well as some paracrine/endocrine molecules such as secretin, vasoactive intestinal polypeptide (VIP) and angiotensin II (Leung & Ip, 2006). These secretagogues act on their respective receptors located at the basolateral plasma membrane of acinar cells, thus evoking downstream signalling transduction pathways mainly in the two forms of intracellular signalling pathways.

ACh and CCK exert their effects on the exocrine pancreas by binding their respective muscarinic receptors and CCK_B receptors, respectively. Both muscarinic and CCK_B receptors in the exocrine pancreas are linked to $G\alpha_q$ G-proteins, use the phospholipase C (PLC)/Ca^{2+} signal transduction pathway, and lead to increased enzyme secretion from acinar cells. On the other hand, the actions of VIP and secretin are mediated via their respective receptors and both activate $G\alpha_s$ G-proteins which stimulates adenylate cyclase (AC), leading to the production of cAMP and the activation of protein kinase A (PKA). PKA and PKC finally trigger ZG fusion with the apical plasma membrane, leading to the polarized secretion of the digestive enzymes or exocytosis into the acinar lumen (Wäsle & Edwardson, 2002).

It is generally accepted that exocytosis is activated by a rise in cytosolic Ca^{2+} levels. Given that the pancreatic acinar cell is not electrically excitable, though, it cannot rely on the extracellular Ca^{2+} influx to stimulate enzyme secretion (Petersen, 1992). The Ca^{2+} concentration required can be achieved from intracellular stores. This concept was well illustrated by experimental data showing that the initial secretory response to either ACh or CCK is independent of extracellular Ca^{2+}. Meanwhile, prolonged secretion does need a surge in extracellular Ca^{2+} as intracellular Ca^{2+} stores are depleted (Petersen & Ueda, 1976). ACh and CCK can evoke various types of Ca^{2+} signals within acinar cells; the deviation is dependent on the concentration of the secretagogue that the acinar cells encounter (Scheele et al., 1987). Low concentrations of ACh or CCK induce weak local Ca^{2+} signals near the secretory pole of the acinar cells. In contrast, a high concentration of secretagogue

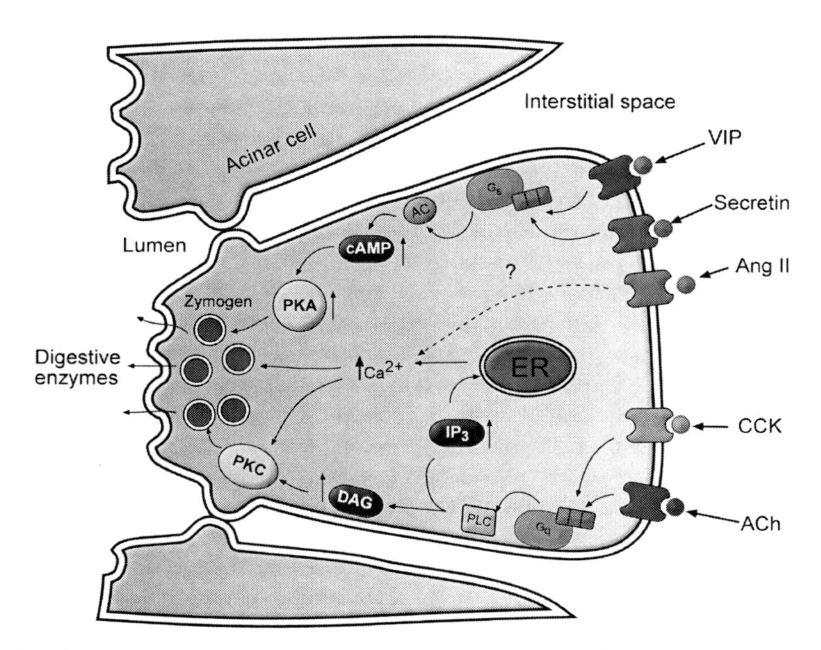

Fig. 2.2 Regulated secretion of digestive enzyme from the acinar cell. The pancreatic acinar cell has two common signaling pathways for the exocytosis of secretory zymogen granules into the acinar lumen. They are the ACh and CCK-mediated PKC/Ca^{2+} pathway and VIP and secretin-mediated cAMP/PKA signaling pathways. The angiotensin II-mediated transduction pathway, probably via Ca^{2+}, has yet to be determined

produces a global Ca^{2+} wave that spreads over the entire cell (Kasai et al., 1993). In addition, secretin and VIP activate an elevation of cAMP levels, and thus activation of PKA (Wäsle & Edwardson, 2002). On the other hand, a local renin-angiotensin system (RAS) demonstrated to exist in the acinar cell may have a functional role in the regulation of protein secretion by pancreatic acinar cells (Leung & Carlsson, 2001). In this context, angiotensin II was shown to stimulate exocytosis from the pancreatic acinar cells in a dose dependent manner, probably through mediation of intracellular Ca^{2+} concentration (Tsang et al., 2004). The localization, expression, regulation and function of the RAS in the exocrine pancreas and its potential roles in pancreatic inflammation will be further discussed in Chapters 6 and 10. Figure 2.2 summarizes some important neurohormonal regulators and their respective receptor-mediated signalling pathways that have stimulatory effects on protein secretion from the pancreatic acinar cell.

2.3 Pancreatic Duct Cells

The pancreatic duct cells represent a minority of the exocrine portion of the pancreas. Nevertheless, these cells are indispensable for normal functioning of the digestive enzymes secreted by acinar cells and integrity of the duodenal mucosa.

Pancreatic duct cells are critical because they are responsible for the secretion of the sodium bicarbonate (HCO_3^-)-rich alkaline fluid that neutralizes gastric chyme that is emptied into the duodenum. The physiological function of bicarbonate is to provide an optimal pH for pancreatic digestive enzymes and to prevent injury to the duodenal mucosa or peptic ulcer disease. Deficiency of the duct cells thus causes maldigestion and malabsorption. While the acinar cell enzymatic secretion has been studied extensively, the underlying mechanism of ductal secretion is relatively less well understood. Duct cells constitute only 10% of the total number of cells and 5% of the total mass of the human pancreas (Githens, 1988). The duct endings are structurally linked with acinar cells via the centroacinar cells, which have several ductal characteristics and are regarded as the terminals of the ductal tree. Digestive enzymes produced by acinar cells are firstly emptied into the intercalated ducts, followed by intralobular ducts, and finally an interlobular duct. Intralobular ducts merge into the main pancreatic duct, which shares the duodenal opening with the common bile duct at the ampulla of Vater (For the structure and relationship between the duct and acinar cells, please see Fig. 1.1. Chapter 1). There is a sphincter made up of smooth muscle, called the sphincter of Oddi, and its contraction simultaneously regulates bile and pancreatic juices entering into the duodenum. Cells in the intercalated and intralobular ducts are the major producers of HCO_3^-. These ducts are lined with principal cells that possess small amounts of rough endoplasmic reticulum, Golgi apparatus and secretory vesicles. Abundant mitochondria are found in the principal cells, reflecting the huge energy demands of the secretory duct cells. The apical membranes of the principal cells possess microvilli while the basal membranes of the principal cells are joined by tight junctions or adherent junctions. Being the largest branches of the network, the interlobular duct cells become columnar in shape and intermingle with goblet cells, which are responsible for mucous production.

The pancreatic juice is a clear HCO_3^- rich alkaline and isotonic fluid. About 1–2 l of pancreatic juice is produced each day in the adult human body. The concentration of HCO_3^- in pancreatic juice is extraordinarily high at 120–140 mmol/l (Domschke et al., 1977), which is fivefold higher than that in plasma. Pancreatic ductal HCO_3^- and acinar enzymatic secretion are increased following consumption of a meal. The primary stimulus for HCO_3^- secretion by ductal epithelium is regulated by the peptide hormone secretin, secreted by the gut endocrine cells (D-cells), in response to gastric acidic chyme, which enters the duodenum, particularly with a pH below 3.5 (Bayliss & Starling, 1902). Fatty acids and high concentrations of bile salts can also induce the production of secretin by the endocrine D-cells located in duodenal mucosa (Hanssen, 1980). Apart from secretin, multiple stimulatory and inhibitory factors, such as CCK and vagal stimulation, are also involved in the regulation of the pancreatic ductal secretion; this multitude of interacting neurohormonal factors is indicative of the complexity in ductal secretion control (Pandol, 2004). In term of molecular and cellular level, secretion of high concentrations of HCO_3^- into the ductal fluid is under the direct control of several ion channels, notably the apical chloride channel called cystic fibrosis transmembrane conductance regulator (CFTR), specifically expressed on the ductal cells. Figure 2.3 presents a schematic

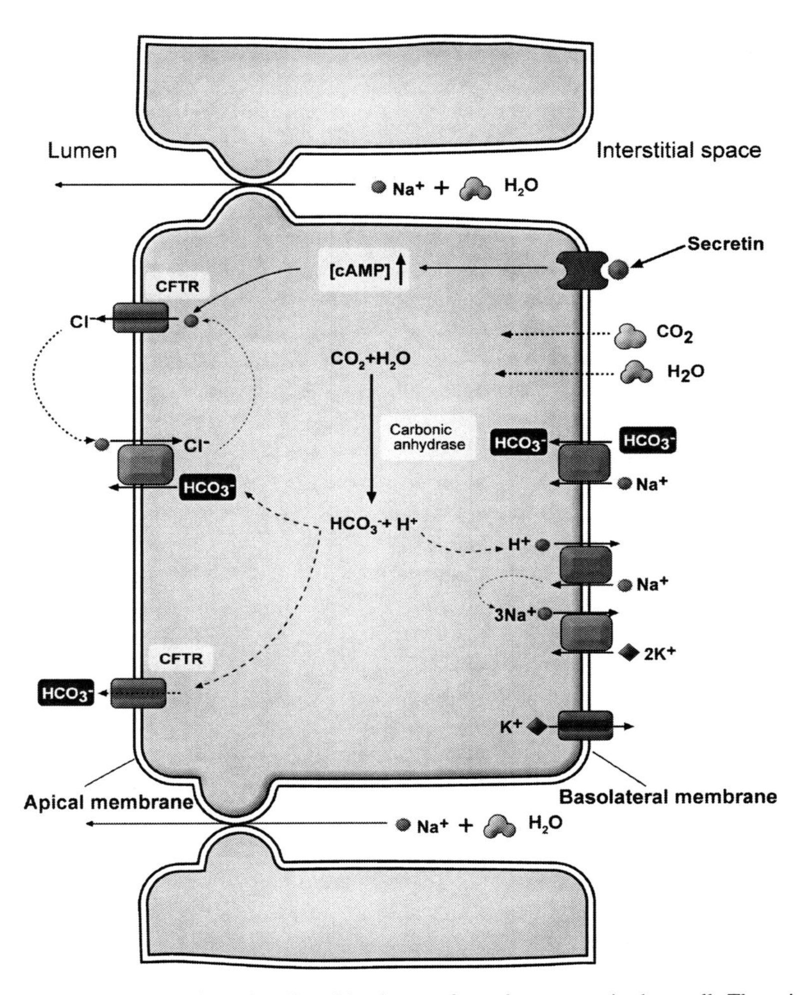

Fig. 2.3 Regulated secretion of sodium bicarbonate from the pancreatic duct cell. The primary stimulus secretin is released by the duodenal endocrine cells in response to the arrival of gastric chyme. Stimulation of the apical membrane-bound CFTR channel is dependent on the elevation of cAMP levels which, in turn, rests on the activation of secretin receptor located at the basolateral membrane of pancreatic duct cell

of the currently accepted cell model for CFTR-mediated bicarbonate secretion by duct cells. In this model, CO_2 enters the cell basolaterally by passive diffusion and carbonic anhydrate, which is expressed within the duct cell, catalyzes the formation of HCO_3^- from CO_2 and H_2O while concomitantly producing protons or H^+. The protons are transported out of the cell through a Na^+/H^+ exchanger on the basolateral membrane. The 2 $K^+/3Na^+$-ATPase then pumps Na^+ out of the cell, thus providing a driving force for the Na^+/H^+ exchanger with the establishment of a concentration gradient of Na^+ ions. Meanwhile, HCO_3^- is secreted out in exchange for Cl^- by a Cl^-/HCO_3^- exchanger located in the apical membrane (Melvin et al., 1999). Importantly, the tight coupling of Cl^- and HCO_3^- secretion depends critically on the regulation and activity of the CFTR channel (Poulsen et al., 1994). Indeed, the CFTR activity or its opening is cAMP dependent which, in turn, is determined by the hormone secretin bound to its receptor located on the basolateral membrane of duct cell. CFTR may be "switched on" for HCO_3^- transportation when the luminal Cl^- is reduced to about 20 mmol/l (Shcheynikov et al., 2006). The clinical relevance of the CFTR is manifested by a genetic defect of this anion channel in patients with cystic fibrosis (see Section 3.2, Chapter 3).

Apart from its established pancreatic secretory function, the pancreatic ductal cell has recently been proposed to be a source of potential progenitors or stem cells. It has been found that the duct cells possess some proliferative capacity for the neogenesis of new ducts, as well as islet cells, even after the morphogenesis of pancreas has been fully completed (Bonner-Weir et al., 2004). An increase in the mitotic activity of the duct cells is obvious when the pancreas is subjected to various degrees of injury, such as duct ligation or streptozotocin-induced destruction of β-cells (Bonner-Weir & Sharma, 2002). Co-expression of several ductal markers and developmental markers, notably PDX1 (a specific pancreatic cell marker), in a subset of duct cells were identified. These findings prompt us to suggest that pancreatic ductal cells may serve as progenitors or stem cells for the generation of new pancreatic cells (Bonner-Weir and Sharma, 2002). Accumulated evidence has further shown that in vitro culture of enriched duct cells from human pancreas or non-obese diabetic (NOD) mice are capable of generating insulin-secreting cells (Bonner-Weir & Sharma, 2002). Moreover, ductal cells have been found to initiate endocrine cell differentiation with the inclusion of several specific transcriptional factors (Heremans et al., 2002). Owing to the lack of specific markers for pancreatic ductal cells, the regenerative capacity of the duct cells remains inconclusive. Further information about pancreatic stem cells, with particular focus on the novel roles of the RAS on their proliferation, differentiation and maturation will be discussed in Chapter 9.

2.4 Pancreatic Islet Cells

In contrast with the massive exocrine portion of the pancreas, the endocrine portion represents only 1–2% of the total mass of the organ. Despite their minority, islet cells mediate indispensable functions in glucose homeostasis. The richly vascularized islets receive about 10–15% of the total pancreatic blood flow despite of their small cell population. Amongst the four major types of endocrine cells, β-cells are dominant, constituting about 80% of the total population of islet cells. The four islet cell types are arranged in a well-defined pattern, such that β-cells are located in the centre surrounded by α- and δ-cells in islet periphery. Blood flows from the centre of the islet to the periphery; this unique arrangement provides a paracrine cell–cell interaction within the islets (see Fig. 1.2, Chapter 1). By virtue of this arrangement, α-cells and δ-cells are exposed to high concentrations of insulin so as to finely control glucagon and somatostatin release, respectively. β- and δ-cells are distributed throughout the pancreas; α-cells are located exclusively in the tail, body and superior part of the head of the pancreas; F-cells or PP-cells that produce pancreatic polypeptide are present in the middle and inferior parts of pancreas head.

The function of pancreatic β-cells is to synthesize and release insulin, which is well known to be an anabolic peptide hormone. It decreases plasma glucose levels by enhancing glucose uptake by peripheral tissues such as adipose and skeletal and suppressing hepatic glucose production. Pre(pro)insulin is cleaved into (pro)insulin during its insertion into the endoplasmic reticulum. (Pro)insulin consists of an amino-terminal β-chain and a carboxy-terminal α-chain as well as a connecting peptide, known as the C-peptide, between the chains. C-peptide enables proper folding of the insulin molecule and formation of disulphide linkages between the α-chain and β-chain. In the endoplasmic reticulum, (pro)insulin is cleaved by a specific endopeptidase known as prohormone convertase into mature insulin. Removal of the C-peptide exposes the end of the insulin chain and thus allows its interaction with insulin receptors (Steiner and Rubenstein, 1997). The free C-peptide and mature insulin are packed into secretory granules in the Golgi apparatus of β-cells and are released into the blood by exocytosis. Only ~5% of the granules are readily releasable; most of the granules (>95%) are reserved and require further chemical modification to become readily releasable. Thus, only a small proportion of insulin can be released by the β-cells under maximal stimulation.

2.4.1 Beta-Cells

Beta-cells are best known for producing insulin in response to changes in plasma levels of major nutrients, particularly glucose, amino acids, and fatty acids. Insulin secretion is also increased by a number of gut hormones such as insulin, glucagon-like peptide-1 (GLP-1), gastric inhibitory peptide (or called glucose-dependent insulinotropic peptide, GIP), secretin and CCK, as well as by vagal and β-adrenergic stimulation. On the other hand, β-cell secretion is decreased by fasting, exercise and somatostatin as well as by enhanced α-adrenergic activity. Beta-cells take up and metabolize glucose, galactose and mannose, and each can provoke insulin secretion by the islet. In this regard, the primary stimulus for insulin release is the nutrient such as amino acid, fatty acids and particularly, glucose. The physiological importance of the primary stimulus glucose can be exemplified by the fundamental concept called, glucose-stimulated insulin secretion (GSIS) (Fig. 2.4). According to this concept, glucose enters the β-cell through a membrane-bound glucose transporter, GLUT2 via facilitative diffusion. The imported glucose is phosphorylated by an enzyme called glucokinase, and a chain reaction of glycolysis ensues, leading to the citric acid cycle in which acetyl-coenzyme A (CoA) and adenosine triphosphate (ATP) are produced. The concomitant increase in the ratio of ATP to ADP inhibits an ATP-sensitive potassium channel so as to reduce K^+ efflux, thereby leading to membrane depolarization of the cell. The depolarization activates a voltage-gated Ca^{2+} channel, which in turn promotes Ca^{2+} influx through the plasma membrane.

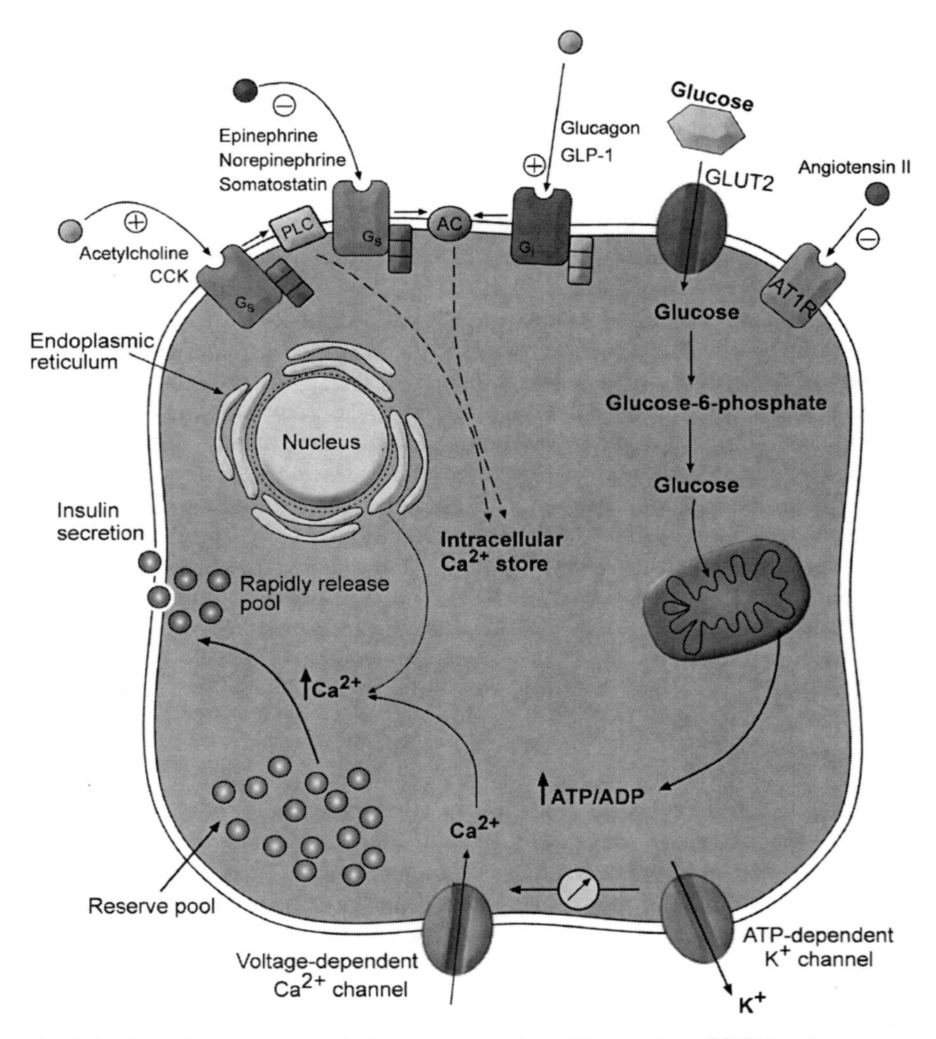

Fig. 2.4 The basic mechanism of glucose-stimulated insulin secretion (GSIS) by the pancreatic beta-cell. The primary stimulus glucose enters the cell via a facilitative GLUT2 transporter and undergoes glucose metabolism which leads to a rise in ATP/ADP ratio, thus inhibiting ATP-sensitive K^+ channel. The depolarization of the K^+ channel activates a voltage-gated Ca^{2+} channel which, in turn, triggers a surge of intracellular Ca^{2+} and finally leads to exocytosis and release of insulin into the blood

A rise in intracellular Ca^{2+} levels finally triggers exocytosis and release of insulin from the secretory granules into the blood circulation (Schuit et al., 2001).

GSIS is well known to exhibit a biphasic action in response to glucose stimulation. Upon glucose challenge, an acute phase of insulin release is evoked, the first-phase insulin secretion, which is characterized by a rapid peak of insulin release and a return to the baseline level. The first-phase of insulin release is followed by a chronic or sustained phase or second-phase insulin secretion, which slowly reaches a plateau and sustained for the duration of the glucose challenge. The rapid first phase is due to the release of insulin stores of presynthesized insulin while the slow second phase is attributed to the release of newly manufactured insulin or insulin biosynthesis (Maechler et al., 2006). On the other hand, amylin is a peptide produced in beta-cells together with insulin in a ratio of one to one that is thought to regulate gastric motility, renal resorption, and metabolic actions. Formation of islet amyloid deposits from amylin reduces beta-cell mass and insulin secretion, thus acting a key pathophysiological factor in the pathogenesis of diabetes (Wookey et al., 2006).

2.4.2 Alpha-Cells

Alpha-cells produce glucagon, which acts in opposition to insulin. In contrast to insulin, its major physiological effect is to increase plasma glucose levels as a catabolic hormone by stimulating de novo synthesis of hepatic glucose via gluconeogenesis. Glucagon secretion is activated by hypoglycemic but inhibited by hyperglycaemic conditions. High plasma levels of amino acids, epinephrine, and vagal activation also stimulate glucagon release. In addition, somatostatin inhibits glucagon release in a paracrine fashion. In general, glucagon can counteract the effect of insulin on glucose homeostasis through its binding to the G_S protein coupled glucagon receptor. Interestingly, it has recently been reported that α-cells can produce a ligand of the growth hormone secretagogue receptor (GHSR), which stimulates the release of growth hormone from the anterior pituitary through a G-protein coupled GHSR (Date et al., 2002). This 28-amino acid peptide called ghrelin was first identified in the stomach (Kojima et al., 1999). Ghrelin stimulates appetite and food intake, while enhancing fat mass deposition and weight gain via its actions on multiple organ systems including the central nervous and gastro-intestinal systems (De Vriese & Delporte, 2008). Recently, it has been proposed that ghrelin-secreting cell called epsilon-cells are identified in the developing and adult human pancreas (Andralojc et al., 2009), indicating that epsilon-cells are ontogenetically and morphogenetically distinct from alpha-cells and beta-cells (see Section 1.2, Chapter 1). In the pancreas, ghrelin has biological effects on islet cell insulin secretion (Date et al., 2002) and acinar enzyme secretion (Zhang et al., 2001) from the endocrine and exocrine pancreas, respectively. In this regard, a locally expressed ghrelin system has been reported in the acinar cells as having a role in the regulation of acinar cell function (Lai et al., 2005). Hence, the acinar cell ghrelin system is subjected to regulation by physiological and pathophysiological stimuli such as gastric acid inhibition, acute pancreatitis, and starvation, suggesting that it is involved in exocrine pancreatic function and dysfunction (Lai et al., 2007).

2.4.3 Delta-Cells and PP-Cells

Somatostatin was originally discovered as a neuropeptide produced by the hypothalamic cell, based on its inhibitory action on the release of growth hormone from the pituitary gland; it was later found in various cells, including the intestinal endocrine and nerve cells as well as in the pancreas. Apart from its classical actions on the "hypothalamus-pituitary axis", somatotstain exhibits a wide spectrum of secretory and motor functions on such organs of the gastrointestinal system as the stomach and intestine as well as the exocrine and endocirne pancreas. There are two biologically active forms of somatostatin synthesized from the precursor molecule, preprosomatostatin, called SS-14 and SS-28 consisting of 14 amino acids and 28 amino acids, respectively. In the intestine, SS-14 was identified in both enteric neurons and enteroendocrine cells whereas SS-28 was reported only in endocrine D-cells (Francis et al., 1990). Somatostatin secretion is regulated by nutrients, gastrointestinal hormones, glucagon, and neurotransmitters, and inhibited by insulin. Physiologically speaking, somatostatin is a profound inhibitor of insulin and glucagon secretion as well as a potent inhibitor of acid secretion by the parietal cells of stomach. In general, administration of somatostatin decreases the assimilation rate of all nutrients from the gastrointestinal system (Unger et al., 1978). In the pancreas, somatostatin is produced by the delta-cells of the endocrine islet cells which contributes a relatively few cell population of the islet (about 5%); however, it has profound biological and physiological effects on the endocrine as well as exocrine pancreas. For example, an inhibitory effect of somatostatin on pancreatic exocrine secretion, particularly digestive enzyme release via its mediation of cholecystokinin and secretin, has been previously recognized in humans (Boden et al., 1975, Schlegel et al., 1977). As far as endocrine pancreas is concerned, somatostatin plays profound and inhibitory effects on insulin and glucagon secretion by the islet cells, thus markedly decreasing blood glucose levels in humans and diabetic subjects (Wahren & Felig, 1976). Recently, angiotensing II type 2 receptors (AT2R) have been identified to be co-localized with somatostatin-producing delta-cells in the endocrine pancreas; interestingly, angiotensin II can stimulate a biphasic release of somatostatin in a dose-dependent manner, probably via mediation of the AT_2 receptors expressed in δ-cells (Wong & Cheung, 2004).

Pancreatic peptide (PP) is produced by a small population of PP-cells or F-cells in the islets. There are only a few F-cells (less than 1% of islet cell population) scattered throughout the islets, primarily concentrated in the dorsal part of the head of the pancreas in a distribution opposite to that of glucagon. PP is a 36-amino acid peptide with a distinctive C-terminal tyrosine amide residue which belongs to the neuropeptide Y and peptide YY family. Its secretion is increased by food ingestion, mainly due to protein and fat content, cholinergic stimulation, and hypoglycaemia, and it is decreased by glucose. Basically, the release of PP is highly dependent upon cholinergic stimulation although some hormonal agents released by meals also regulate postprandial release. Although receptors of PP have been found in gut-brain axis, the physiological function of PP remains obscure. Among the putative roles, PP is known to have diverse effects on the gastrointestinal secretion and motility such as inhibition of exocrine and exocrine pancreatic secretion, stimulation of gastric acid secretion, relaxation of gallbladder motility, and regulation of migrating motor complex of gut smooth muscle (Podolsky, 1994). More recently, it has been reported that elevated plasma levels of PP are associated with increased intra-abdominal fat accumulation and thus insulin resistance in humans (Tong et al., 2007).

References

Andralojc KM, Mercalli A, Nowak KW, Albarello L, Calcagno R, Luzi L, Bonifacio E, Doglioni C and Piemonti L. Ghrelin-producing epsilon cells in the developing and adult human pancreas. *Diabetologia* **52**:486–493, 2009.

Bayliss WM and Starling EH. The mechanism of pancreatic secretion. *J Physiol* **28**:325–353, 1902.

Beger HG. The pancreas: an integrated textbook of basic science, medicine, and surgery. Blackwell, Oxford, 2008.

Berry SM and Fink AS. Insulin inhibits secretin-stimulated pancreatic bicarbonate output by a dose-dependent neurally mediated mechanism. *Am J Physiol* **270**:G163–G170, 1996.

Bolender RP. Stereological analysis of the guinea pig pancreas. I. Analytical model and quantitative description of nonstimulated pancreatic exocrine cells. *J Cell Biol* **61**:269–287, 1974.

Boden G, Sivitz MC, Owen OE, Essa-Koumar N, Landon JH. Somatostatin suppresses secretin and pancreatic exocrine secretion. *Science* **190**:163–165, 1975.

Bonner-Weir S and Sharma A. Pancreatic stem cells. *J Pathol* **197**:519–526, 2002.

Bonner-Weir S, Toschi E, Inada A, Reitz P, Fonseca SY, Aye T and Sharma A. The pancreatic ductal epithelium serves as a potential pool of progenitor cells. *Pediatr Diabetes* **5**:16–22, 2004.

Chey WY and Chang T. Neural hormonal regulation of exocrine pancreatic secretion. *Pancreatology* **1**:320–335, 2001.

Chey WY, Shay H and Shuman CR. External pancreatic secretion in diabetes mellitus. *Ann Intern Med* **59**:812–821, 1963.

Date Y, Nakazato M, Hashiguchi S, Dezaki K, Mondal MS, Hosoda H, Kojima M, Kangawa K, Arima T, Matsuo H, Yada T and Matsukura S. Ghrelin is present in pancreatic alpha-cells of humans and rats and stimulates insulin secretion. *Diabetes* **51**:124–129, 2002.

De Vriese C and Delporte C. Ghrelin: a new peptide regulating growth hormone release and food intake. *Int J Biochem Cell Biol* **40**:1420–1424, 2008.

Domschke S, Domschke W, Rösch W, Konturek SJ, Sprügel W, Mitznegg P, Wünsch E and Demling L. Inhibition by somatostatin of secretin-stimulated pancreatic secretion in man: a study with pure pancreatic juice. *Scand J Gastroenterol* **12**:59–63, 1977.

Francis B, Baskin D, Saunders D and Ensinck J. Distribution of somatostatin-14 and somatostatin-28 gastrointestinal-pancreatic cells of rats and humans. *Gastroenterology* **99**:1283–1291, 1990.

Githens S. The pancreatic duct cell: proliferative capabilities, specific characteristics, metaplasia, isolation, and culture. *J Pediatr Gastroenterol Nutr* **7**:486–506, 1988.

Hanssen LE. Pure synthetic bile salts release immunoreactive secretin in man. *Scand J Gastroenterol* **15**:461–463, 1980

Hardt PD, Krauss A, Bretz L, Porsch-Ozcürümez M, Schnell-Kretschmer H, Mäser E, Bretzel RG, Zekhorn T and Klör HU. Pancreatic exocrine function in patients with type 1 and type 2 diabetes mellitus. *Acta Diabetol* **37**:105–110, 2000.

Heremans Y, Van De Casteele M, in't Veld P, Gradwohl G, Serup P, Madsen O, Pipeleers D and Heimberg H. Recapitulation of embryonic neuroendocrine differentiation in adult human pancreatic duct cells expressing neurogenin 3. *J Cell Biol* **159**:303–312, 2002.

Horiuchi A, Iwatsuki K, Ren LM, Kuroda T and Chiba S. Dual actions of glucagon: direct stimulation and indirect inhibition of dog pancreatic secretion. *Eur J Pharmacol* **237**:23–30, 1993.

Howard-McNatt M, Simon T, Wang Y and Fink AS. Insulin inhibits secretin-induced pancreatic bicarbonate output via cholinergic mechanisms. *Pancreas* **24**:380–385, 2002.

Iwatsuki N and Petersen OH. Electrical coupling and uncoupling of exocrine acinar cells. *J Cell Biol* **79**:533–545, 1978.

Kasai H, Li YX and Miyashita Y. Subcellular distribution of Ca2+ release channels underlying Ca2+ waves and oscillations in exocrine pancreas. *Cell* **74**:669–677, 1993.

Kojima N, Hosoda H, Date Y, Nakazato M, Matsuo H and Kangawa K. Ghrelin is a growth hormone-releasing acylated peptide from stomach. *Nature* **402**:565–660, 1999.

Korc M, Owerbach D, Quinto C and Rutter WJ. Pancreatic islet-acinar cell interaction: amylase messenger RNA levels are determined by insulin. *Science* **213**:351–353, 1981.

Lai KC, Cheng CHK, Ko WH and Leung PS. Ghrelin system in pancreatic AR42J cells: its ligand stimulation evokes calcium signaling through ghrelin receptors. *Int J Biochem Cell Biol* **37**:887–900, 2005.

Lai KC, Cheng CHK and Leung PS. The ghrelin system in acinar cells: localization, expression, and regulation in the exocrine pancreas. *Pancreas* **35**:e1–e8, 2007.

Leung PS and Carlsson PO. Tissue renin-angiotensin system: its expression, localization, regulation and potential role in the pancreas. *J Mol Endocrinol* **26**:155–164, 2001.

Leung PS and Ip SP. Pancreatic acinar cell: its role in acute pancreatitis. *Int J Biochem Cell Biol* **38**:1024–1030, 2006.

Maechler P, Carobbio S and Rubi B. In beta-cells, mitochondria integrate and generate metabolic signals controlling insulin secretion. *Int J Biochem Cell Biol* **38**:696–709, 2006.

Melvin JE, Park K, Richardson L, Schultheis PJ and Shull GE. Mouse down-regulated in adenoma (DRA) is an intestinal Cl^-/HCO_3^- exchanger and is up-regulated in colon of mice lacking the NHE3 Na^+/H^+ exchanger. *J Biol Chem* **274**:22855–22861, 1999.

Murakami T, Fujita T, Taguchi T, Nonaka Y and Orita K. The blood vascular bed of the human pancreas, with special reference to the insulo-acinar portal system. canning electron microscopy of corrosion casts. *Arch Histol Cytol* **55**:381–395, 1992.

Nakagawa A, Stagner JI and Samols E. Suppressive role of the islet-acinar axis in the perfused rat pancreas. *Gastroenterology* **105**:868–875, 1993.

Okabayashi Y, Maddux BA, McDonald AR, Logsdon CD, Williams JA and Goldfine ID. Mechanisms of insulin-induced insulin-receptor downregulation. Decrease of receptor biosynthesis and mRNA levels. *Diabetes* **38**:182–187, 1989.

Pandol SJ. Neurohumoral control of exocrine pancreatic secretion. *Curr Opin Gastroenterol* **20**:435–438, 2004.

Patel R, Singh J, Yago MD, Vilchez JR, Martínez-Victoria E and Mañas M. Effect of insulin on exocrine pancreatic secretion in healthy and diabetic anaesthetised rats. *Mol Cell Biochem* **261**:105–110, 2004.

Petersen OH. Human Physiology. Blackwell, Oxford, 2007.

Petersen OH. Stimulus-secretion coupling: cytoplasmic calcium signals and the control of ion channels in exocrine acinar cells. *J Physiol* **448**:1–5, 1992.

Petersen OH and Ueda N. Pancreatic acinar cells: the role of calcium in stimulus-secretion coupling. *J Physiol* **254**:583–606, 1976.

Podolsky DK. Peptide growth factors in the gastrointestinal tract. In LR Johnson (ed), Physiology of the gastrointestinal tract, vol 1, 3rd edn. Raven Press, New York, NY, pp 1–128, 1994.

Poulsen JH, Fischer H, Illek B and Machen TE. Bicarbonate conductance and pH regulatory capability of cystic fibrosis transmembrane conductance regulator. *Proc Natl Acad Sci USA* **91**:5340–5344, 1994

Saito A, Williams JA and Kanno T. Potentiation of cholecystokinin-induced exocrine secretion by both exogenous and endogenous insulin in isolated and perfused rat pancreata. *J Clin Invest* **65**:777–782, 1980.

Scheele G, Adler G and Kern H. Exocytosis occurs at the lateral plasma membrane of the pancreatic acinar cell during supramaximal secretagogue stimulation. *Gastroenterology* **92**:345–353, 1987.

Schlegel W, Raptis S, Harvey RF, Oliver JM, Pfeiffer EF. Inhibition of cholecystokinin-pancreaozymin release by somatostatin. *Lancet* **2**:166–168, 1977.

Schuit FC, Huypens P, Heimberg H and Pipeleers DG. Glucose sensing in pancreatic beta-cells: a model for the study of other glucose-regulated cells in gut, pancreas, and hypothalamus. *Diabetes* **50**:1–11, 2001.

Shcheynikov N, Wang Y, Park M, Ko SB, Dorwart M, Naruse S, Thomas PJ and Muallem S. Coupling modes and stoichiometry of Cl-/HCO3- exchange by slc26a3 and slc26a6. *J Gen Physiol* **127**:511–524, 2006.

Simon T, Marcus A, Royce CL, Chao F, Mendez T and Fink AS. Hyperglycemia alone does not inhibit secretin-induced pancreatic bicarbonate secretion. *Pancreas* **20**:277–281, 2002.

Steiner DF and Rubenstein AH. Proinsulin C-peptide-biological activity. *Science* **277**:531–532, 1997.

Tong J, Utzschneider KM, Carr DB, Zraika S, Udayasankar J, Gerchman F, Knopp RH and Kahn SE. Plasma pancreatic polypeptide levels are associated with differences in body fat distribution in human subjects. *Diabetologia* **50**:439–442, 2007.

Tsang SW, Cheng CHK and Leung PS. The role of pancreatic renin-angiotensin system in acinar digestive enzyme secretion and in acute pancreatitis. *Regul Pept* **119**:213–219, 2004.

von Schönfeld J, Goebell H and Müller MK. The islet-acinar axis of the pancreas. *Int J Pancreatol* **16**:131–140, 1994.

Unger RH, Dobbs RS and Orci L. Insulin, glucagon and somatostatin secretion in the regulation of metabolism. *Annu Rev Physiol* **40**:307–343, 1978.

Wahren J and Felig P. Influence of somatostatin on carbohydrate disposal and absorption in diabetes mellitus. *Lancet* **2**:1213–1216, 1976.

Wäsle B and Edwardson JM. The regulation of exocytosis in the pancreatic acinar cell. *Cell Signal* **14**:191–197, 2002.

Williams JA and Goldfine ID. The insulin-pancreatic acinar axis. *Diabetes* **34**:980–986, 1985.

Wookey PJ, Lutz TA and Andrikopoulos S. Amylin in the periphery: an updated mini-review. *ScientificWorldJournal* **6**:1642–1652, 2006.

Wong PF and Cheung WT. Immunohistochemical colocalization of type II angiotensin receptors with somatostatin in rat pancreas. *Regul Pept* **117**:195–205, 2004.

Zhang W, Chen M, Chen X, Segura BJ and Mulholland MW. Inhibition of pancreatic protein secretion by ghrelin in the rat. *J Physiol* **537**:231–136, 2001.

Chapter 3
Common Pancreatic Disease

3.1 Pancreatitis

Pancreatitis is an inflammatory disease of the pancreas characterized by acute and chronic condition as well as varying duration and severity. Acute pancreatitis (AP) is a severe abdominal inflammation, characterized by parenchymal edema, necrosis with occasional presence of pseudocysts, abscess, hemorrhage, and inflammatory cell infiltration (Chan & Leung, 2007a). Its disease spectrum varies from mild edematous to severe disease with fatal complications. Pancreatic acinar cells are very susceptible to inflammation during an episode of AP attack. It is believed that some initiating factors can lead to the premature transformation of inactive pro-proteases into active proteases within the pancreas. The active proteases degrade a number of cellular proteins such as the structural protein F-actin, thus leading to the collapse and malfunction of acinar and pancreatic damage, a process known as autodigestion of the pancreas (Singh et al., 2001).

3.1.1 Etiology and Prognosis

Some AP-triggering environmental toxins are known to directly affect pancreatic acinar cells, leading to rupture and necrosis, and eventually an extensive inflammatory response. Epidemiological studies have revealed that the morbidity rate of the disease depends on which regions are examined, ranging from 50 to 800 cases per 1,000,000 people annually. Actually, AP is more prevalent in Caucasian than in Asian populations. The incidence rate is 700–800 and 150–420 per million annually in the USA and UK, respectively (Banks, 2002); however, it is just 106–205 per million annually in Japan (Sekimoto et al., 2006). Similar studies in China and Hong Kong have been inconsistent. In 2007, there were only 1976 patients with AP admitted to hospital from 15 Chinese tertiary care centres during the period from 1990 to 2005 (Bai et al., 2007). Despite this discrepancy, there are an increasing number of individuals with AP, probably due to life-style changes and increased exposure to risk factors in recent decades.

P.S. Leung, *The Renin-Angiotensin System: Current Research Progress in The Pancreas*, Advances in Experimental Medicine and Biology 690,
DOI 10.1007/978-90-481-9060-7_3,

The diagnosis of AP entails several characteristic clinical features, such as abdominal pain at the upper epigastric region, vomiting, and elevation of plasma pancreatic enzyme levels, such as α-amylase and lipase. Hyperstimulation or pancreatic lesions might account for leakage of pancreatic enzymes into circulation. It should be emphasized that elevated levels of plasma pancreatic enzymes are a commonly used indicator for AP, though they do not generally reflect the degree of disease severity. It is noted that the elevation of plasma α-amylase tends to be cleared most rapidly from the circulation and it is the first enzyme to fall during resolution of AP. In view of this fact, if a patient with AP presents more than a few days after onset of abdominal pain, serum amylase can be normal or only slightly elevated. In this case, serum lipase is however cleared less rapidly and thus this enzyme is more likely to remain elevated in patients whose onset of symptoms is several days earlier. In addition, serum amylase levels can also be elevated in non-AP conditions such as pancreatic duct obstruction, pancreatic cancer, appendicitis, bowel obstruction and renal failure. On the other hand, measurement of serum C-reactive peptide and interleukin-6 (IL-6) provides additional information, particularly within 48 h of the onset of AP symptoms (Davies & Hagen 1997). For further evaluations, approaches such as ultrasonography and computed tomography imaging are more effective and reliable; these diagnostic procedures offer a better understanding of the cause of the disease, including the presence of gallstones or dilatation of the common bile duct.

In the clinical setting, about 80% of AP can be etiologically determined while the remaining cases are classified as idiopathic or without a known cause. Alcoholism and gallstone obstruction (choledocholithiasis) are the two major causes of the disease, accounting for about 35 and 45% of AP cases, respectively. Other causes of AP, to name but a few, include drug uses, trauma, genetic and infection. The cardinal feature of alcoholism-induced AP is characterized by the formation of proteinaceous plugs in bile ducts thus activating proteolytic enzymes and leading to pancreatic autodigestion. Gallstone obstruction-induced AP is, however, due mainly to gallstone formation which results in blockage of ampulla of Vater and thus reflux of biliary and pancreatic juices into pancreatic ducts, eventually leading to pancreatic inflammation. Different countries and regions have various proportions of pancreatitis caused by alcoholism and gallstones. Alcoholism is a major cause in some regions while gallstones may be a major cause in others. A similar incidence rate between these two etiological factors has also been reported (Gullo et al., 2002). However, it should be noted that the risk of acute alcoholic pancreatitis and acute gallstone-induced pancreatitis is around 2–3% and 0.63–1.48%, respectively, suggesting that these two risk factors can not fully account for the incidence of AP. In fact, idiopathic pancreatitis accounts for the third most common cause of AP clinically. Other risk factors including hyperlipidemia, viral infection (HIV, mumps and hepatitis B), and endoscopic retrograde cholangiopancreatiography (ERCP), as well as surgical procedures and medications may be implicated in the development of AP (Chan & Leung, 2007a).

The AP recurrence rate is relatively high. A cohort study in Sweden reported that about 21% of AP patients suffer a recurrence, with nearly 70% of them exhibiting a second episode within three months of the first episode (Andersson et al., 2004). A similar follow-up study conducted in Japan indicated an overall recurrence rate of 37% (Sekimoto et al., 2006). If choledocholithiasis is not treated properly, the recurrence rate of gallstone-induced pancreatitis could reach 32–61% (Delorio et al., 1995). Patients with alcoholism-induced pancreatitis and idiopathic pancreatitis have been reported to have relapse rates of around 46 and 3.2%, respectively, indicating that recurrence varies in an etiology-dependent manner (Pelli et al., 2000).

Some patients may gradually develop chronic pancreatitis (CP) if recurrent AP is not well controlled. Basic research studies have revealed that repeated episodes of AP result in a progressive development to CP. The so-called "necrosis-fibrosis theory" hypothesizes that residual pancreatic damage, especially necrosis, can gradually lead to parenchymal destruction and fibrosis replacement (Ammann & Muellhaupt, 1994). It is believed that repeated acinar injury leads to activation of pancreatic stellate cells (PSCs). The activated PSCs secrete fibrogenic factors, thus triggering progressive substitution of functional exocrine pancreas and subsequent development of CP (Leung & Chan, 2009). It should be emphasized that AP and CP are two distinct diseases with discrete features in terms of pancreatic morphology and clinical outcome. Patients with AP usually exhibit pancreatic swelling (edematous or necrotic cell death), while the pancreata of CP patients usually undergoes atrophy and apoptosis. In AP, the damage is confined to the exocrine pancreas and exocrine dysfunction leads to pancreatic exocrine insufficiency; meanwhile, CP or repeated episode of AP attack causes endocrine dysfunction that emerges from advanced stage of CP, notably in the form of pancreatic fibrosis which accounts for not only pancreatic exocrine but also endocrine insufficiency, the latter being attributable to impaired glucose tolerance or diabetes.

Mortality is the major complication of severe AP. The overall mortality rate of AP lies between 7.5 and 20%, depending on the diagnostic criteria chosen and the age of the patients. Most patients are susceptible to death within the first few weeks after the onset of symptoms. Early mortality is associated with the development of systemic inflammatory response syndrome (SIRS), aggravated by multiple organ dysfunction syndrome (MODS). In this context, death is generally due to local complications, notably occurrence of infection resulting from pancreatic necrosis. The clinical parameters of MODS include dyspnea, shock, bleeding tendency, and elevated blood urea (Sekimoto et al., 2006). In terms of organ failure, pulmonary dysfunction is commonly observed. Acute respiratory distress syndrome (ARDS) is usually the primary manifestation of AP-associated complications and is believed to trigger MODS (Bosma & Lewis, 2007). On the other hand, the development of ARDS may further exacerbate the severity of pancreatitis since hypoxia is a well-known enhancer of inflammatory responses. Undoubtedly, the more organs that fail, particularly if the liver and kidney are involved, the higher the mortality rate is (McFadden, 1991). The mortality rate in AP-induced ARDS surges when renal failure has ensued. Actually, there is a close association between pancreatic necrosis and development of SIRS as well as MODS. It is believed that necrotizing pancreatitis causes the release of overwhelming pro-inflammatory mediators and, subsequently, over-activates inflammatory responses, leading to SIRS and distant organ lesions and thus death ensues.

3.1.2 Treatments

Although AP has been studied for decades, promising and effective therapy is still not available. In most cases, patients develop mild, self-limiting AP, requiring no special treatment or surgery. Traditional approaches are generally palliative and nonspecific including analgesic (opiate) administration for pain relief, intravenous fluids for volume depletion and nothing by mouth for pancreatic rest. By doing so, the patients may be discharged from the hospital within weeks. However, around 20% of patients develop severe AP, concomitant with lethal complications that bring the mortality rate up to 20–25% in these patients, compared with 1–3% in a mild attack. In such cases, patients should be monitored under intensive care unit and surgery should be justified in such case as gallstone-induced AP.

Enteral/parentenal nutrition, antibiotic treatment, surgical removal of necrotic tissue, and some surgical manipulations such as cholecystectomy are the first-line treatment modalities for clinical AP (Makola et al., 2007). However, many such treatments are rather passive in nature (e.g. enteral/parentenal nutrition) while others are invasive (e.g. surgical management). Most of these treatments are targeted for the complications of the disease (e.g. sepsis or pain management) rather than the primary insults. In view of this fact, clinicians are in search of new effective alternatives for curing AP. Potential therapeutic approaches derived from basic research include, but are not limited to, the protease inhibitor gabexate, the anti-secretagogue agent octreotide, the anti-inflammatory drug lexipafant, and an antioxidant regimen. Unfortunately, thus far, these potential therapeutic methods have yielded unsatisfactory or marginal results and warrant further extensive investigation (Chan & Leung, 2007a; Leung & Chan, 2009).

Emerging data from basic research have shed some lights on potential therapeutic strategies against AP, such as cyclooxygenase (COX)-2 inhibition, substance P antagonism, and heat shock protein (HSP) activation. COX-2 expression has been found to be up-regulated in caerulein-induced pancreatitis in both mice (Song et al., 2002) and rats (Zhou et al., 2004). Treatment with specific COX-2 inhibitors, such as celecoxib and NS-398, could alleviate pancreatic injury (Song et al., 2002). Genetic knock-out of COX-2 leads to resistance against experimental pancreatic injury and its associated lung injury (Song et al., 2002). In addition, the expression of substance P, the physiological agonist of neurokinin 1 receptor (NK1R), has been shown to be upregulated during experimental pancreatitis, thus implicating its role in AP-induced lung injury (Bhatia et al., 1998; Lau et al., 2005a); a pharmacological antagonist for its receptor was also protective against caerulein-induced pancreatitis and associated lung damage (Lau et al., 2005a). On the other hand, the chaperon protein and HSP have also been closely linked with AP pathogenesis; secreatagogue and arginine-induced pancreatitis was shown to provoke pancreatic HSP protein expression and HSP induction was shown to be protective against experimental

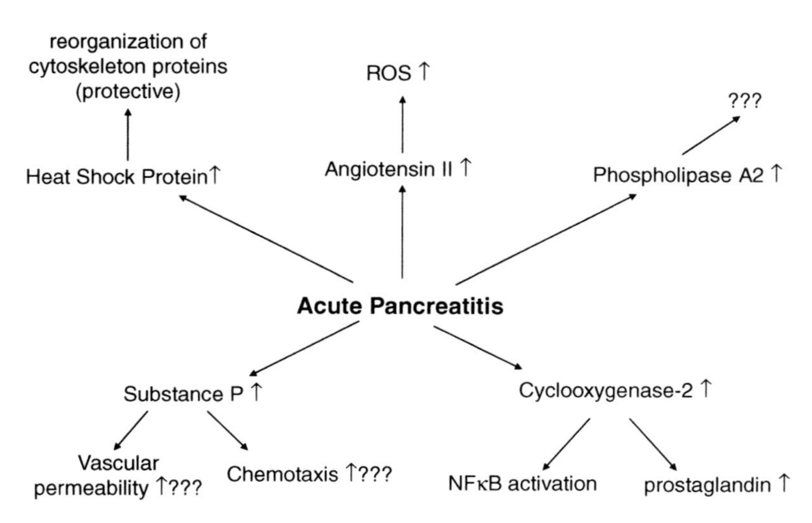

Fig. 3.1 Schematic diagram illustrating the simplified relationship between AP pathophysiology and some novel biological factors (modified from Chan & Leung, 2007a)

pancreatitis (Tashiro et al., 2001). Last but not least, we have shown that angiotensin II, the vasoactive peptide, is crucial in mediating pro-inflammatory responses during AP (Chan & Leung, 2007b). The novel roles of angiotensin II and its type 1 receptor (AT1R) blockade and thus its clinical relevance to pancreatic inflammation will be further discussed in Chapter 10. Potential pathogenetic factors of acute pancreatitis implicated and discussed above in the pathogenesis of AP are summarized in Fig. 3.1.

3.2 Cystic Fibrosis

Cystic fibrosis (CF) is an autosomal recessive disease commonly found about once in every 25,000 births in the Caucasian population while it occurs rare in native Africans and Asians. The disease is a result of a genetic defect leading to abnormal sodium chloride and water movement across the epithelial membranes. Where such membranes line the pancreatic duct and small respiratory airways, they produce dense and viscous mucus that cannot be properly cleared, thus causing pancreatic insufficiency and pulmonary disorder. The genetic defect involves a particular membrane protein expressed in the epithelial cells, called cystic fibrosis transmembrane conductance regulator (CFTR). Mutations in the CFTR are categorized into five groups according to their repercussion on CFTR protein synthesis and its chloride channel function. Class I or stop codon mutations result in truncated non-functional CFTR; Class II mutations consist of aberrantly folded CFTR protein that is degraded by the cell quality control system; Class III mutations lead to defective regulation of the CFTR protein; Class IV mutations cause defective chloride conductance; and

Class V mutations interfere with normal transcription, thereby reducing the amount of normal CFTR (Proesmans et al., 2008).

3.2.1 Etiology and Prognosis

One of the genetic mutations of CF gene is due to a single mutation (amino acid 508) located on the long arm of chromosome 7 that codes for the CFTR protein. This genetic defect is characterized by abnormalities in exocrine gland function that result in altered ion composition and increased viscosity of epithelial cell secretions. The affected tissues include the secretory cells of the sweat glands, salivary glands, small intestine, lungs, vas deferens, and the exocrine pancreas (Voynow & Scanlin, 2005). CF is manifested by elevated electrolyte concentrations, reduced ion permeability, and impaired luminal ion secretion upon stimulation by secretagogues (Tucker et al., 2003). Thus, the defect in perturbed salt and water transport leads to secondary alterations of pancreatic, intestinal and pulmonary functions. Pancreatic exocrine insufficiency is clinically obvious in 85–90% of CF patients (Baker et al., 2005).

As mentioned, CFTR is expressed in various cell types, including the pancreas. It is comprised of two membrane-spanning domains and two nucleotide-binding domains separated by a regulatory R domain. The two membrane-spanning domains form a low-conductance chloride channel pore. It is regulated by ATP binding and hydrolysis at the nucleotide-binding domains following initial phosphorylation of the R domain. CFTR functions as a chloride channel in the apical membrane of epithelial cells that regulate ion transport (Schwiebert et al., 1999). It has inhibitory effects on apical Na^+ permeability across epithelia while activating non-CFTR chloride channels. Owing to the widespread presence of CFTR throughout the body, CF is a multisystem disorder affecting many organs, especially the gastrointestinal tract, pancreas and lungs.

Failure of the function of CFTR results in impaired chloride transport at the apical surface of epithelial cells and dysregulation of other transporters, such as the chloride-coupled bicarbonate transport and sodium channel activity (Reddy et al., 1999; Choi et al., 2001). The aqueous medium of the intestinal lumen becomes hyper-viscous in CP patients due to excessive and abnormal intestinal mucoprotein (Forstner et al., 1984). In the pancreas, these viscous secretions cause luminal obstruction of pancreatic ducts, leading to acinar cell destruction, fibrosis, and ultimately pancreatic exocrine insufficiency which is characterized by a decrease or absence of digestive enzymes and concomitant with maldigestion and malabsorption of nutrients. In view of the close interaction between exocrine and endocrine pancreas, patients with CF usually develop with pancreatic exocrine along with endocrine insufficiency as manifested by other pancreatic conditions such as acute and chronic AP, pancreatic surgery and pancreatic cancer (Czako et al., 2009). The pathophysiology of lung disease in CF is more complex than that of pancreatic disease. A major finding is that the airway mucus is thick and viscous as a result of insufficient fluid secretion into the airway. The lung epithelium secretes fluid in a mechanism that requires CFTR while absorbing fluid that requires apical ENaC sodium channels. In patients with CF, the reduced activity of CFTR shifts the balance more toward absorption, and a thick mucous layer is generated that inhibits the ciliary clearance of foreign bodies. The result is an increased rate and severity of infections and thus inflammatory processes that contributes to the destructive process in the lung (Ingbar et al., 2009). In contrast, the levels of salivary secretion are normal or raised while lingual lipase levels are elevated (Guy-Crotte et al., 1996). Although CF is the most common lethal genetic disease in Caucasians, there is often a delay between the onset of symptoms and definitive diagnosis. Blood spot CF screening tests can be performed in newborn babies. A definitive diagnostic sweat test will then be given if the screening suggests CF. About one in five people with CF are diagnosed at birth when their gut is blocked by extra thick meconium.

3.2.2 Treatments

The conventional treatment of CF is usually palliative, alleviating signs and symptoms and treating organ dysfunction, including replacement of pancreatic enzymes, vitamin and nutritional supplements, airway clearance techniques, daily physiotherapy and antibiotics for pulmonary infections (Davis et al., 1996; Ramsey, 1996).

Despite advances in treatment, there remains no cure for CF. It has become apparent that there is a need for a more effective and convenient therapy. New therapies directed at the basic defect represent the only potential approach to truly treating CF. The identification of the gene responsible for CF (CFTR protein), in 1989, provided insight into a potential treatment for CF. Gene therapy, the transfer of a normal copy of the CFTR gene into the lungs of CF patients, has been proposed as an attractive alternative to the conventional approach (Riordan, 2008; Ratjen, 2007). Genes are most commonly transferred into cells through viruses and liposomes. Viruses have evolved to enter the cells of the body efficiently. Scientists have harnessed this property for gene therapy by inserting a copy of the therapeutic gene of interest into the virus, which then directs it into the cell. The adenovirus is an example of a virus that has been used extensively for gene therapy. On the other hand, liposomes are fatty substances that naturally adhere to the surface of cells, thus facilitating entry into cells. The CF gene could potentially be transferred into the airway cells; however, considerable challenges lie ahead, particularly with regard to efficiency of gene transfer and persistence of transferred gene expression (Atkinson, 2008). The field is moving forward rapidly, particularly pertaining to the development of better virus and liposome vectors. Apart from the aforementioned state-of-the-art methodology, there are a number of potential drugs which are emerging for the treatment of CF and are in clinical trials. They include some anti-bacterial formulations, anti-inflammatory agents, ion channel modulating agents, and agents that correct the underlying gene defect (Jones & Helm, 2009). The pathogenetic pathways and common symptoms of CF are briefly summarized in Fig. 3.2.

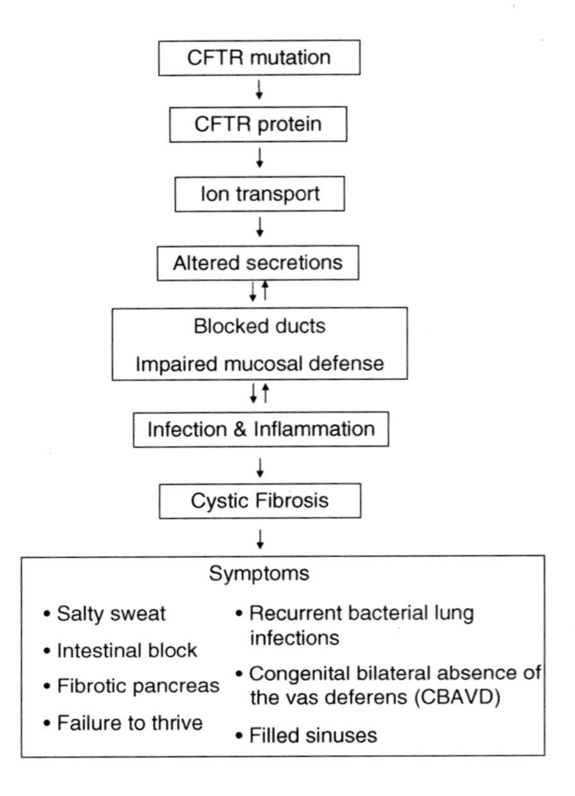

Fig. 3.2 The pathogenesis of cystic fibrosis disease and its common signs and symptoms

3.3 Pancreatic Cancer

3.3.1 Etiology and Prognosis

Pancreatic cancer is a deadly malignant disease with an extremely high mortality rate in humans. It is classified into tumours arising from the pancreatic exocrine tissue (ductal adenocarcinoma) and pancreatic endocrine tissue (insulinoma, glucagonoma, gastrinoma and VIPoma), as well as atypical neoplasms (lymphomas, mesotheliomas and sarcomas) (Hruban & Adsay, 2009). It is the fourth leading cause of cancer in the United States, with an estimated 37,680 new cases and 34,290 deaths in 2008 (Jemal et al., 2008). Ductal adenocarcinoma accounts for approximately 80% of all pancreatic cancers (Lewis, 2006). Only 23% of patients with pancreatic cancer of exocrine origin will survive for 1 year, while about 4% will survive for 5 years. In contrast, pancreatic endocrine tumours are relatively rare accounting for less than 2% of pancreatic neoplasms. In this section, we thus focus on the ductal adenocarcinoma as an example of pancreatic cancer for further discussion below.

The risk of developing pancreatic cancer is much higher beyond the age over 50 and is more prevalent in males. Risk factors include family history, cigarette smoking, a diet rich in fat and meat, and exposure to chemicals. In addition, patients with diabetes mellitus, chronic pancreatitis and gastric surgery are more susceptible to the disease (Dunphy, 2008). Diagnosis of pancreatic cancer in an early stage is extremely difficult, mainly due to the nonspecific symptoms and the lack of accurate diagnostic tests. Hence, more than 80% of patients with pancreatic cancer are diagnosed only in advanced or late tumour stages (stage II–IV), often when the tumours metastasize without a recourse of tumour resection (Canto, 2008). Furthermore, the propensity for tumour metastasis and intrinsic resistance to conventional chemo-radiation treatments renders pancreatic cancer very difficult to handle clinically. Surgical resection of localized tumours is the only potentially curative option available for pancreatic cancer patients (Ozawa et al., 2001). Unfortunately, pancreatic cancer is a highly aggressive cancer and, with the late diagnosis that characterizes many cases, only about 10–15% of patients are considered suitable for the surgical option. There are two standard resection procedures, including resection of the pancreatic head and of the left-side pancreatic tissues. In the case of a local advanced tumour, a total pancreatectomy may be chosen for particular patients (Loos et al., 2008a). Pancreatic surgery has significantly improved and the associated postoperative morbidity and mortality have decreased markedly over the last three decades. Despite these advances, the prognosis for patients with pancreatic cancer remains poor and this poor prognosis is the reason for pancreatic cancer's high mortality rate.

3.3.2 Treatments

Complete surgical resection of the tumour is only possible in 10–25% of patients with pancreatic cancer, while about half of them are subjected to adjuvant therapies. Chemotherapy with 5-fluorouracil (5-FU), cisplatin, oxaliplatin, mitomycin C, and doxorubicin in mono- or combination treatments are the major options for pancreatic cancer. In the case of locally unresectable or metastatic adenocarcinoma, chemotherapy was found to enhance overall 1-year survival, but not to appreciable improve long-term (5-year) survival (Loos et al., 2008b). In 1996, gemcitabine was approved as the standard chemotherapeutic agent in monotherapy or combination therapy for locally advanced and metastatic pancreatic cancer (Barton-Burke, 1999; Hui & Reitz, 1997). Gemcitabine has been reported to improve survival and alleviate disease-related symptoms (Almhanna & Kim, 2008). Single chemotherapy with gemcitabine is considered the standard for patients with advanced pancreatic cancer along with other drugs including 5-FU. Meanwhile, targeted therapies inhibiting specific pathways for the growth and progression of malignant tumors are under intensive investigations. The epidermal growth factor receptor (EGFR) mediated pathway is one of the most promising targets for the treatment. In over 90% of cases, EGFR is over-expressed to stimulate pancreatic tumour growth (Lemoine et al., 1992). Indeed, the EGFR tyrosine inhibitor erlotinib has been shown to increase median survival and, in 2005, the U.S. FDA approved the use of erlotinib in combination with gemcitabine for treating locally advanced, unresectable, or metastatic pancreatic cancer (Danovi et al., 2008). Gene therapy using antisense oligonucleotides or ribozymes could also correct a genetic defect or mutation. For example, K-ras and p53 tumour suppressor gene mutations are found in about 70%

of pancreatic cancer and have been predominant targets in preclinical gene therapy studies (Danovi et al., 2008).

Despite these recent advances, treatment of pancreatic cancer remains difficult and relatively ineffective due to its intrinsic broad resistance to cytotoxic drugs. Some recent studies have focused on the transcription factor nuclear factor kappa B (NF-κB), which is proposed to be a central determinant in the induction and manifestation of chemo-resistance in pancreatic cancer cells. Since constitutive activation of NF-κB is a hallmark of pancreatic cancer accounting for profound chemoresistance, suppression of NF-κB activity might be a useful strategy for increasing sensitivity towards cytostatic drug treatment (Sebens et al., 2008). However, solid tumors apparently exhibit comprehensive protection from induction of apoptosis, so pharmacological inhibition of NF-κB only appears to be insufficient. Thus, NF-κB inhibition should be used as a chemo-sensitizing adjuvant in combination with other cytostatic drugs.

Several new immunotherapies have been developed for potential treatment of pancreatic cancer. Chemotherapy, which reduces tumour cells by preferentially poisoning proliferating cells, is not a cancer cure per se. Immunotherapy, however, could potentially eliminate quiescent tumour (stem) cells. Although immunotherapy is in the early stage of development and has not yet replaced conventional chemotherapy, it may in the future be an additional approach in fighting cancer and eradicating tumour stem cells (Schulenburg et al., 2006). Table 3.1 summarizes the different forms of pancreatic tumours that arise from exocrine and endocrine origins, together with their characteristics and the current treatments.

3.3.3 Alternative Approach Using Traditional Chinese Medicine

Based on herbal medicine knowledge that has accrued over centuries, traditional Chinese medicine provides a potential approach for alternative treatments to pancreatic cancer. Chinese herbal extracts have been extensively investigated in various clinical trials of anti-tumour studies, including studies of leukemia, lung, ovarian, breast and pancreatic cancers. Indeed, a number of Chinese herbs have been identified as having anti-cancer properties (Wang et al., 2005). For example, Curcumin has been reported to possess a wide range of beneficial properties, including anti-inflammatory, antioxidant, chemopreventive and chemo-therapeutic activities. Curcumin influences multiple signalling pathways, including the following: cell survival pathways regulated by NF-κB, Akt, and growth factors; Nrf2-dependent cytoprotective pathways; and matrix metallo-proteinase-dependent metastatic and angiogenic pathways (Park et al., 2005; Zheng & Chen, 2004). Additionally, the traditional Chinese medicinal *Brucea javanica* was recently shown to exhibit various biological activities such as anti-malarial (O'Neill et al., 1987), anti-inflammatory (Hall et al., 1983), and hypoglycaemic effects (NoorShahida et al., 2009), thus implicating its clinical use for the treatment of various diseases.

Brucea javanica is a shrub of about 3-meter height. The fruit of *Brucea javanica*, or Ya Dan Zi in Chinese (meaning crow bile fruit), contains oils (glyceroltrileate,

Table 3.1 A brief summary of classification, characteristics and current treatments of exocrine and endocrine tumors of the pancreas

Origin	Types	Characteristics	Treatments
Pancreatic exocrine tumors	Adenocarcinoma	The most common type of pancreatic cancer, accounting for 75% of all pancreas cancer; nearly all of these are ductal adenocarcinoma; cause back pain when tumor grow large and invade nerves	Surgical resection: pancreaticoduodenectom, total pancreatectomy and distal pancreatectomy
	Acinar cell carcinoma	Rare cancerous tumor produces excessive amounts of digestive enzymes. Unusual skin rashes, joint pain and increased increased eosinophils level	First-line chemotherapy: Gemcitabine
	Adenosquamous carcinoma	Similar to adenocarcinoma that it forms glands, but it flattens as it grows. It can mimic other types of cancer that show squamous differentiation	Second line chemotherapy: 5-flourouracil (5-Fu), irinotecan, celecoxib, cisplatin and oxaliplatin
	Giant cell tumor	Extremely rare and is not aggressive as adenocarcinomas. It has unusually large cells	Combined therapy: ICM-C225 + Gem, erlotinib + Gem
	Intraductal papillary-mucinous neoplasm (IPMN)	Rare but very distinctive tumor. It grows along the pancreatic duct and appears to be a fingerlike projection into the duct	
	Mucinous cystade-nocarcinoma	Rare, cystic, fluid-containing pancreas tumor and can develop into cancer over time. The space within the spongy tumor is filled with a think fluid called mucin	Radiation therapy and chemoradiotherapy
	Pancreatoblastoma	Rare malignant tumor occurs primarily in children, and called pancreatic cancer of infancy	
Pancreatic endocrine tumor	Insulinoma	Produce large amounts of insulin which result in hypoglycemia	Surgical resection Chemotherapy: Streptozocin, dacarbazine, doxorubicin and 5-Fu Combination therapy

Table 3.1 (continued)

Origin	Types	Characteristics	Treatments
	Glucagonoma	Produce excessive amounts of glucagon which result in severe dermatitis, mild diabetes, stomatitis, anemia, and weight loss	Distal pancreatectomy Standard chemotherapy: streptozocin and dacarbazine Octreotide: Reduce elevated glucagon levels, and control the hyperglycemia and dermatitis
	Gastrinoma	Release large quantities of the hormone gastrin into the blood stream leading to severe duodenal ulcers and persistent diarrhea	Surgical resection Chemotherapy: Proton pump inhibitors, like lansoprazole, pantoprazole esomeprasole, in high doses to control hypersecretion of gastric acid
	VIPoma	Releasing large amounts of the hormone VIP into the blood stream. Symptoms include watery diarrhea, hypokalemia, and either achlorhydria or hypochlor-hydria.	Surgical excision Chemotherapy: Octreotide to reduce circulating VIP levels and control diarrhea No specific chemotherapy for VIPoma patients
	Somatostatinoma	Less common, releasing large quantities of the hormone somatostatin into the blood stream	Combination treatment with intravenous 5-FU and streptozotocin or doxorubicin and 5-FU

Fig. 3.3 Photographs of *Brucea javanica*. (**a**) The plant. (**b**) The fruit. (**c**) The dry fruit. (**d**) Chemical structure of brucein D

oleic acid, linoleic acid), alkaloids, quassinoids, bursatol, brucein A, B, C, D, E, F and H etc (Fig. 3.3). In term of cancer, extracts or compounds from *Brucea javanica*, particularly brucein D, have anti-proliferative and cytotoxic effects on a number of cancer cell types; they include, but are not limited to, the lung, liver, breast and oesophageal cancers (Lau et al., 2005b). In our laboratory, we have recently performed preliminary screening tests on nine commonly used Chinese herbal medicines for anti-pancreatic cancer activities. Among these, an ethanolic extract of *Brucea javanica* fruit has also been identified to possess potent cytotoxicity that induced marked apoptosis in several human pancreatic cancer cell lines, including PANC-1, SW1990 and CAPAN-1 while exhibiting less cytotoxic action on Hs68 cells (a human foreskin fibroblast cell line); specifically, it produced chromatin condensation and fragmentation as well as activation of the

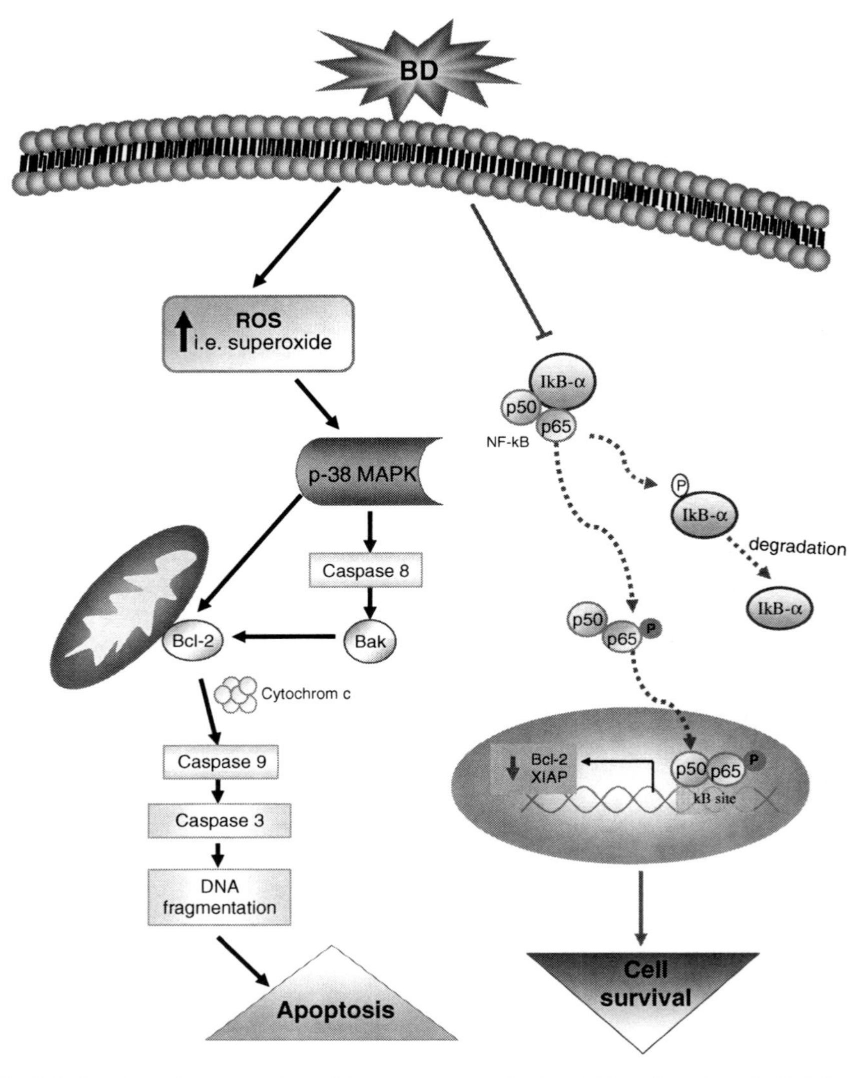

Fig. 3.4 A schematic presentation of the proposed mechanism (s) by which brucein D induces apoptosis and anti-apoptosis in PANC-1 cells

proteolytic cleavage of caspase 3 in these human pancreatic cancer cell lines (Lau et al., 2008). Furthermore, brucein D, a quassinoid compound found abundantly in *Brucea javanica* fruit, displayed potent anti-proliferative activity and induced apoptosis in PANC-1 cells through activation of the p-38 mitogen-activated protein kinase (MAPK) signal transduction pathway (Lau et al., 2009). Both intrinsic and extrinsic apoptotic pathways are activated following p38-MAPK phosphorylation and, ultimately, the effector caspase 3 is activated, causing irreversible apoptotic cell death in pancreatic cancer cells. On the other hand, Brucein D treatment depletes intracellular glutathione levels in PANC-1 cells, while triggering activation of the NADPH oxidase isoforms $p22^{phox}$ and $p67^{phox}$, and increasing superoxide generation. Although intracellular reactive oxygen species and NADPH oxidase activities can be inhibited by the presence of an antioxidant (e.g. N-acetyl cysteine), inhibition of NF-κB activity in brucein D-treated cells appears to be independent of ROS changes (Lau et al., 2010). These results suggest that brucein D elicits apoptosis in pancreatic cancer cells by activating the redox-regulated p38 MAPK signalling pathway and reducing anti-apoptotic activity through inhibition of NF-κB activation.

Taken the above substantial data together, these findings provide experimental evidence that supports the potential use of traditional Chinese medicinal such as *Brucea javanica* fruit extract, in the treatment of pancreatic cancer and, more importantly, brucein D is a promising candidate compound for further development into a mechanism-driven anti-pancreatic cancer agent. A schematic presentation of the proposed mechanism(s) by which brucein D induces MAPK-mediated apoptosis and NF-κB-mediated survival in PANC-1 cells is presented in Fig. 3.4.

3.4 Diabetes Mellitus

3.4.1 Etiology and Prognosis

Blood glucose concentration has a normal range of 3.8–7 mmol/l (70–125 mg/dl). When the body is not able to maintain normal glucose status, glucose cannot enter into cells to supply them with energy. Prolonged high blood glucose levels, a state known as hyperglycemia, ultimately leads to diabetes mellitus. The American Diabetes Association (ADA, 2009) distinguishes among normoglycemia, impaired glucose tolerance (IGT, also known as pre-diabetes), and overt diabetes mellitus based on fasting plasma glucose (FPG) levels and 2-h oral glucose tolerance test (OGTT) results as follows: normoglycemia is FPG <6.0 mmol/l (110 mg/dl) and 2-h OGTT <7.7 mmol/l (140 mg/dl); IGT is FPG levels ≥ 6.0 mmol/l (110 mg/dl) but <7.0 mmol/l (125 mg/dl), and 2-h OGTT ≥ 7.8 mmol/l (140 mg/dl) but <11.1 mmol/l (200 mg/dl); and overt diabetes mellitus is FPG ≥ 7.0 mmol/l (125 mg/dl) and 2-h OGTT >11.1 mmol/dl (200 mg/dl). Individuals with impaired FPG or IGT are generally euglycemic and have normal or near normal glycated hemoglobin levels (HbA_{1c}, normal range 3.5–5.5%), but may progress to overt diabetes without treatment. Diabetes is often concomitant with chronic complications such as nephropathy, retinopathy and neuropathy which in turn can lead to end-stage renal failure, blindness and diabetic foot, respectively.

Diabetes mellitus is classified as type 1 (T1DM), type 2 (T2DM), gestational diabetes (GDM), and other specific types, based on its specific pathogenesis (American Diabetes Association, 2009). According to the definition of the World Health Organization (WHO), the two major types, namely T1DM and T2DM, account for approximately 10 and 90% of diabetics, respectively. T1DM (also previously known as insulin-dependent or juvenile diabetes) results from an absolute deficiency of insulin secretion from pancreatic β-cells. Most of the T1DM incidence is due to immune-mediated disease with autoimmune features, such as the presence of islet cell antibodies (e.g., islet cell antigen, islet cell surface antigen, GAD65 and GAD67). T1DM pathogenesis is thought to involve both genetic susceptibility and environmental factors. T2DM results from relative insulin deficiency and insulin resistance (also previously known as non-insulin dependent or adult-onset diabetes). While the etiology of T2DM is multifactorial, it is associated with strong genetic and environmental factors (e.g., single-diabetes gene, diet, physical inactivity and obesity). In the former case, one form of T2DM called maturity onset diabetes of the young (MODY) has been attributed to dysfunction of a single gene. MODY is an autosomal dominant genetic disease which leads to metabolic abnormality, notably defective insulin secretion. Based on the influence of specific single gene concerned, it is divided into various forms, such as MODY1, MODY2, MODY3, MODY 4 etc, depending on which particular gene is involved. For example, if the defective gene occurs in the hepatocyte nuclear factor 4α (HNF 4α) and pancreatic duodenal homeobox factor-1 (PDX-1), they are called MODY 1 and MODY 4, respectively (Table 3.2).

Some confusion has arisen from the above classification scheme. For example, some patients with severe T2DM may also require insulin therapy in order to maintain euglycemic control. In addition, age is no longer a classification criterion. Though most commonly observed before the age of 30, T1DM can develop at any age (Eisenbarth, 2007). Meanwhile, T2DM once supposed to develop later in life, can also occur in young people (Song & Hardisty, 2008). In any case, loss of beta-cell mass over time is a common feature of both T1DM and T2DM, leading to impaired glucose tolerance, hyperglycemia and overt diabetes (Butler et al., 2003). On the other hand, GDM is defined as any degree of glucose intolerance with an onset during pregnancy (American Diabetes Association, 2009). The cause of GDM remains ambiguous, but placental hormones produced during pregnancy may reduce sensitivity to insulin. Repeated GDM over multiple pregnancies may increase the incidence of permanent diabetes. On the other hand, the remaining ~1% of cases, termed secondary diabetes, are considered complications of other conditions or medications. Secondary diabetes is usually temporary, unless due to total pancreatectomy. Conditions associated with secondary diabetes include disorders that impair pancreatic functions, such as hemochromatosis, chronic pancreatitis, cystic fibrosis, and pancreatic trauma, as well as drugs such as beta blockers, glucocorticoids, antiretrovirals, opiates, and alcohol. Table 3.3 summarizes some cardinal features and that distinguish T1DM from T2DM.

Table 3.2 Some examples of single recessive dominant genes as exemplified by maturity onset diabetes of the young (MODY) which is associated with metabolic syndrome disorder as observed in T2DM

Gene	Type of MODY	Genetic defect	Metabolic disorder	Clinical features
Hepatocyte nuclear factor 4α	MODY 1	Autosomal dominant	Impaired insulin secretion	Rare and progressive form of early onset disease
Glucokinase	MODY 2	Autosomal dominant	Impaired insulin secretion	Mild and relatively stable early onset disease
Hepatocyte nuclear factor 1α	MODY 3	Autosomal dominant	Impaired insulin secretion	Progressive form of early onset disease
Pancreatic duodenal homeobox factor-1	MODY 4	Autosomal dominant	Impaired insulin secretion	Early onset disease
Mitochondrial DNA	–	Maternal DNA	Impaired insulin secretion	Diabetes associated with deafness
Insulin	–	Autosomal dominant	Defective insulin production	Very rare
Insulin receptor	–	Autosomal dominant or recessive	Impaired insulin signaling pathway	Severe insulin resistance

Table 3.3 Some key features in term of etiology, metabolic feature and clinical presentation that distinguish T1DM from T2DM

Type	*T1DM*	*T2DM*
Etiology		
• Autoimmune features	Yes (e.g. islet cell antigens, GAD65 & 67)	No
	Moderate	Very strong (e.g. MODY)
• Genetic susceptibility	Viruses, diet and stress	Physical inactivity
• Environmental factors		
Metabolic features		
• Insulin deficiency	Severe	Moderate
• Insulin resistance	Moderate	Severe
• Diabetic ketoacidosis	Yes	No
Clinical presentation		
• Age of onset	Most occur < 40 years old	Most occur > 40 years
• Body weight	Commonly decrease	Sometimes decrease
• Body mass index	Mostly <25	Mostly >25

3.4.2 Treatments

The incidence of diabetes is ever increasing globally in terms of mortality and morbidity (Wild et al., 2004). Unfortunately, adequate preventive and curative management for diabetes is lacking. Most treatments are intended to slow or prevent disease progression and complications (Reimann et al., 2008). Ideally, T1DM treatment should be focused on β-cell regeneration and early intervention (Meier, 2008). While T1DM may be incurable, its onset and progression may be delayed. Insulin therapy remains the mainstay treatment of T1DM since glycemic control is crucial for preventing chronic complications. Hypoglycemia remains the major disadvantage of insulin injection treatment. Physiologically, prevention of β-cell loss would be a more direct treatment. In this context, recent studies have proposed novel approaches, such as immunoregulatory techniques, to deal with autoimmune β-cell destruction (Phillips et al., 2008). In addition, pancreatic or renal-pancreatic transplantation together with long-term immunosuppressive drugs can be an affective treatment course for T1DM (Stratta & Alloway, 1998). Although islet β-cell transplantation has been performed experimentally in animals and human patients, its major obstacle and practicability for clinical use are due to the very limited availability of islets (Shapiro et al., 2006). In order to resolve this issue, a stem cell-based approach to islet cell transplantation is promising and may someday circumvent this problem as well as providing a cure but not only treatment for T1DM. In this regard, hyperglycemia has been completely reversed in some cases as reported recently (Voltarelli et al., 2007). Additional information on current research concerning pancreatic stem cells, with particular focus of the novel roles of RAS in this area, will be discussed in Chapter 9.

As far as T2DM is concerned, improvements of insulin secretion and/or insulin sensitivity along with reduction of chronic complications are the major considerations in T2DM treatment (Stumvoll et al., 2005). Key preventive strategies include primary prevention, decreasing risk factors, and enhancing β-cell function and cell mass (Reimann et al., 2008). Life-style changes, such as reducing obesity and increasing physical activity can reduce the risk of developing T2DM and related complications (Pan et al., 1997; Gillies et al., 2007). Pharmacological interventions, such as oral hypoglycemic drugs, are the most effective means of therapy, when diet, exercise and change of life style are not sufficient (Torgerson et al., 2004). There are three major categories of oral anti-diabetic drugs currently available based on the two primary causes of T2DM: insulin sensitizers that enhance insulin sensitivity of peripheral tissues at the levels of skeletal muscle, adipose tissue and liver; β-cell secretagogues that improve β-cell secretion, function and cell mass; and α-glucosidase inhibitors that delay or inhibit carbohydrate absorption from the intestine. Apart from these classes of drugs, some novel treatment modalities and novel candidate drugs such as combination drugs and RAS blockers, respectively, the later being fully discussed in Chapter 8, provide significant and effective approaches for treating T2DM. Current treatments of oral hypoglycaemic drugs and some combination therapy for T2DM, and their proposed mechanisms involved are briefly tabulated in Table 3.4.

Table 3.4 Currently and potentially used oral hypoglycemic medications and proposed mechanisms for the management of T2DM

Major class of drugs	Site of actions and proposed mechanisms	Some specific drugs
β-cell secretagogues		
Sulfonylureas	• Act on the pancreas • Enhance β-cell secretion by acting on the ATP-dependent potassium channel • Use with or without insulin but hypoglycemia	– Tolbutamide Glipizide Meglitinides Nateglinide – Exenatide, Liraglutide; – Sitagliptin, Vidagliptin;
GLP-1 analogues/DPP-IV inhibitors/Amylin analogue	• Act on the pancreas • Enhance β-cell secretion and/or cell mass • Do not cause weight gain but GI side effects	– Synthetic amylin
α-glucosidase inhibitors	• Act on the gastrointestinal tract • Lower blood glucose by delaying the digestion and absorption of carbohydrates • Do not cause weight gain and hypoglycemia but concomitant with gas, bloating and diarrhea	– Acarbose – Miglitol
Insulin sensitizers		
Biguanides	• Act primarily on the liver • Decrease liver's glucose production and slightly increase muscle glucose uptake • Do not cause weight gain and hypoglycemia while inducing nausea, diarrhea or loss of appetite	– Metformin – Metformin (extended release) – Metformin (liquid)
Thiazolidinediones (TZDs)	• Act on the peripheral tissues • Decrease insulin resistance at the muscle & liver levels • Improve cholesterol and triglyceride status while but cause weight gain	– Pioglitazone – Rosiglitazone – Troglitazone
RAS blockers	• Act on the pancreas and the peripheral tissues • Improve β-cell function, structure and/or insulin resistance	– ACEIs (Ramipril) – ARBs (Losartan, Valsartan)

Table 3.4 (continued)

Major class of drugs	Site of actions and proposed mechanisms	Some specific drugs
Combination drugs	• Act on different tissue-organ levels • Potential advantages and disadvantages for each drug in the combination listed separately above	– Metformin + TZD, – Metformin + DPP-IV inhibitor, – TZD + Sulfonylurea

RAS, Renin-angiotensin system; ACEIs, Angiotensin-converting enzyme inhibitors; ARBs, Angiotensin receptor blockers; GLP-1, Glucagon-like peptide-1; DPP-IV, Dipeptidyl peptidase IV.

References

Almhanna K and Kim R. Second-line therapy for gemcitabine-refractory pancreatic cancer: is there a standard? *Oncology* **22**:1176–1183, 2008.

American Diabetes Association. Diagnosis and classification of diabetes mellitus. *Diabetes Care* **32**(Suppl 1):S62–S67, 2009.

Ammann RW and Muellhaupt B. Progression of alcoholic acute to chronic pancreatitis. *Gut* **35**:552–556, 1994.

Andersson R, Andersson B, Haraldsen P, Drewsen G and Eckerwall G. Incidence, management and recurrence rate of acute pancreatitis. *Scand J Gastroenterol* **39**:891–894, 2004.

Atkinson TJ. Cystic fibrosis, vector-mediated gene therapy and relevance of toll-like receptors: a review of problems, progress, and possibilities. *Curr Gene Ther* **8**:201–207, 2008.

Bai Y, Liu Y, Jia L, Jiang H, Ji M, Huang K, Zou X, Li Y, Tang C, Guo X, Peng X, Fang D, Wang B, Yang B, Wang L and Li Z. Severe acute pancreatitis in China: etiology and mortality in 1976 patients. *Pancreas* **35**:232–237, 2007.

Baker SS, Borowitz D and Baker RD. Pancreatic exocrine function in patients with cystic fibrosis. *Curr Gastroenterol Rep* **7**:227–233, 2005.

Banks PA. Epidemiology, natural history, and predictors of disease outcome in acute and chronic pancreatitis. *Gastrointest Endosc* **56**:S226–S230, 2002.

Barton-Burke M. Gemcitabine: a pharmacologic and clinical overview. *Cancer Nurs* **22**:176–183, 1999.

Bhatia M, Saluja AK, Hofbauer B, Frossard JL, Lee HS, Castagliuolo I, Wang CC, Gerard N, Pothoulakis C and Steer ML. Role of substance P and the neurokinin 1 receptor in acute pancreatitis and pancreatitis-associated lung injury. *Proc Natl Acad Sci USA* **95**:4760–4765, 1998.

Bosma KJ and Lewis JF. Emerging therapies for treatment of acute lung injury and acute respiratory distress syndrome. *Expert Opin Emerg Drugs* **12**:461–77, 2007.

Butler AE, Janson J, Bonner-Weir S, Ritzel R, Rizza RA and Butler PC. Beta-cell deficit and increased beta-cell apoptosis in humans with type 2 diabetes. *Diabetes* **52**:102–110, 2003.

Canto MI. Screening and surveillance approaches in familial pancreatic cancer. *Gastrointest Endosc Clin N Am* **18**:535–553, 2008.

Chan YC and Leung PS. Acute pancreatitis: animal models and recent advances in basic research. *Pancreas* **34**:1–14, 2007a.

Chan YC and Leung PS. Angiotensin II type 1 receptor-dependent nuclear factor-kappaB activation-mediated proinflammatory actions in a rat model of obstructive acute pancreatitis. *J Pharmacol Exp Ther* **323**:10–18, 2007b.

Choi JY, Muallem D, Kiselyov K, Lee MG, Thomas PJ and Muallem S. Aberrant CFTR-dependent HCO_3^- transport in mutations associated with cystic fibrosis. *Nature* **410**:94–97, 2001.

Czako L, Hegyi P, Rakonczay Z, Wittmann T and Otsuki M. Interactions between the endocrine and exocrine pancreas and their clinical relevance. *Pancreatology* **9**:351–359, 2009.

Danovi SA, Wong HH and Lemoine NR. Targeted therapies for pancreatic cancer. *Br Med Bull* **87**:97–130, 2008.

Davies MG and Hagen PO. Systemic inflammatory response syndrome. *Br J Surg* **84**:920–935, 1997.

Davis PB, Drumm M and Konstan MW. Cystic fibrosis. *Am J Respir Crit Care Med* **154**: 1229–1256, 1996.

Delorio AV Jr, Vitale GC, Reynolds M and Larson GM. Acute biliary pancreatitis. The roles of laparoscopic cholecystectomy and endoscopic retrograde cholangiopancreatography. *Surg Endosc* **9**:392–396, 1995.

Eisenbarth GS. Update in type 1 diabetes. *J Clin Endocrinol Metab* **92**:2403–2407, 2007.

Dunphy EP. Pancreatic cancer: a review and update. *Clin J Oncol Nurs* **12**:735–741, 2008.

Forstner J, Wesley A, Mantle M, Kopelman H, Man D and Forstner G. Abnormal mucus: nominated but not yet elected. *J Pediatr Gastroenterol Nutr* **3**(Suppl 1):S67–S73, 1984.

Gillies CL, Abrams KR, Lambert PC, Cooper NJ, Sutton AJ, Hsu RT and Khunti K. Pharmacological and lifestyle interventions to prevent or delay type 2 diabetes in people with impaired glucose tolerance: systematic review and meta-analysis. *BMJ* **334**:299, 2007.

Gullo L, Migliori M, Oláh A, Farkas G, Levy P, Arvanitakis C, Lankisch P and Beger H. Acute pancreatitis in five European countries: etiology and mortality. *Pancreas* **24**:223–227, 2002.

Guy-Crotte O, Carrere J and Figarella C. Exocrine pancreatic function in cystic fibrosis. *Eur J Gastroenterol Hepatol* **8**:755–759, 1996.

Hall IH, Lee KH, Imakura Y, Okano M and Johnson A. Anti-inflammatory agents. III. Structure-activity relationships of brusatol and related quassinoids. *J Pharm Sci* **72**:1282–1284, 1983.

Hruban RH and Adsay NV. Molecular classification of neoplasms of the pancreas. *Hum Pathol* **40**:612–623, 2009.

Hui YF and Reitz J. Gemcitabine: a cytidine analogue active against solid tumors. *Am J Health Syst Pharm* **54**:162–170, 1997.

Ingbar DH, Bhargava M and O'Grady SM. Mechanisms of alveolar epithelial chloride absorption. *Am J Physiol* **297**:L813–L815, 2009.

Jemal A, Siegel R, Ward E, Hao Y, Xu J, Murray T and Thun MJ. Cancer statistics, 2007. *CA Cancer J Clin* **58**:71–96, 2008.

Jones AM and Helm JM. Emerging treatments in cystic fibrosis. *Drugs* **69**:1903–1910, 2009.

Lau HY, Wong FL and Bhatia M. A key role of neurokinin 1 receptors in acute pancreatitis and associated lung injury. *Biochem Biophys Res Commun* **327**:509–515, 2005a.

Lau FY, Chui CH, Gambari R, Kok SH, Kan KL, Cheng GY, Wong RS, Teo IT, Cheng CH, Wan TS, Chan AS and Tang JC. Antiproliferative and apoptosis-inducing activity of *Brucea javanica* extract on human carcinoma cells. *Int J Mol Med* **16**:1157–1162, 2005b.

Lau ST, Lin ZX and Leung PS. Role of reactive oxygen species in brucein D-mediated p38 mitogen-activated protein kinase and nuclear factor κB signaling pathways in human pancreatic adenocarcinoma cells. *Br J Cancer* **102**:583–593, 2010.

Lau ST, Lin ZX, Liao YH, Zhao M, Cheng CH and Leung PS. Brucein D induces apoptosis in pancreatic adenocarcinoma cell line PANC-1 through the activation of p38-mitogen activated protein kinase. *Cancer Letters* **281**:42–52, 2009.

Lau ST, Lin ZX, Zhao M and Leung PS. Brucea javanica fruit induces cytotoxicity and apoptosis in pancreatic adenocarcinoma cell lines. *Phytother Res* **22**:477–486, 2008.

Lemoine NR, Hughes CM, Barton CM, Poulsom R, Jeffery RE, Kloppel G., Hall PA and Gullick WJ (1992). The epidermal growth factor receptor in human pancreatic cancer. *J Pathol* **166**: 7–12, 1992.

Leung PS and Chan YC. Role of oxidative stress in pancreatic inflammation. *Antioxid Redox Signal* **11**:135–165, 2009.

Lewis BC. Development of the pancreas and pancreatic cancer. *Endocrinol Metab Clin* **35**: 387–404, 2006.

Loos M, Kleeff J, Friess H and Buchler MW. Approaches to localized pancreatic cancer. *Curr Oncol Rep* **10**:212–219, 2008a.

Loos M, Kleeff J, Friess H and Buchler MW. Surgical treatment of pancreatic cancer. *Ann N Y Acad Sci* **1138**:169–180, 2008b.

Makola D, Krenitsky J and Parrish CR. Enteral feeding in acute and chronic pancreatitis. *Gastrointest Endosc Clin N Am* **17**:747–64, 2007.

McFadden DW. Organ failure and multiple organ system failure in pancreatitis. *Pancreas* **6** (Suppl 1):S37–S43, 1991.

Meier JJ. Beta cell mass in diabetes: a realistic therapeutic target? *Diabetologia* **51**:703–713, 2008.

NoorShahida A, Wong TW and Choo CY. Hypoglycemic effect of quassinoids from Brucea javanica (L.) Merr (Simaroubaceae) seeds. *J Ethnopharmacol* **124**:586–591, 2009.

O'Neill MJ, Bray DH, Boardman P, Chan KL, Phillipson JD, Warhurst DC and Peter W. Plants as sources of antimalarial drugs. Part 4. Activity of *Brucea Javanica* fruits against chloroquine-resistant *Plasmodium falciparum* in vitro an against Plasmodium berghei in vivo. *J Nat Prod* **50**:41–48, 1987.

Ozawa F, Friess H, Kunzli B, Shrikhande SV, Otani T, Makuuchi M and Buchler MW. Treatment of pancreatic cancer: the role of surgery. *Dig Dis* **19**:47–56, 2001.

Pan XR, Li GW, Hu YH, Wang JX, Yang WY, An ZX, Hu ZX, Lin J, Xiao JZ, Cao HB, Liu PA, Jiang XG, Jiang YY, Wang JP, Zheng H, Zhang H, Bennett PH and Howard BV. Effects of diet and exercise in preventing NIDDM in people with impaired glucose tolerance. The Da Qing IGT and Diabetes Study. *Diabetes Care* **20**:537–544, 1997.

Park SD, Jung JH, Lee HW, Kwon YM, Chung KH, Kim MG and Kim CH. Zedoariae rhizoma and curcumin inhibits platelet-derived growth factor-induced proliferation of human hepatic myofibroblasts. *Int Immunopharmacol* **5**:555–569, 2005.

Pelli H, Sand J, Laippala P and Nordback I. Long-term follow-up after the first episode of acute alcoholic pancreatitis: time course and risk factors for recurrence. *Scand J Gastroenterol* **35**:552–555, 2000.

Phillips B, Giannoukakis N and Trucco M. Dendritic cell mediated therapy for immunoregulation of type 1 diabetes mellitus. *Pediatr Endocrinol Rev* **5**:873–879, 2008.

Proesmans M, Vermeulen F and De Boeck K. What is new in cystic fibrosis: from treating symptoms to correction of the basic defect. *Eur J Pediatr* **167**:839–849, 2008.

Ramsey BW. Management of pulmonary disease in patients with cystic fibrosis. *N Engl J Med* **335**:179–188, 1996.

Ratjen F. New pulmonary therapies for cystic fibrosis. *Curr Opin Pulm Med* **13**:541–546, 2007.

Reddy MM, Light MJ and Quinton PM. Activation of the epithelial Na+ channel (ENaC) requires CFTR Cl^- channel function. *Nature* **402**:301–304, 1999.

Reimann M, Bonifacio E, Solimena M, Schwarz PE, Ludwig B, Hanefeld M and Bornstein SR. An update on preventive and regenerative therapies in diabetes mellitus. *Pharmacol Ther* **121**: 317–331, 2008.

Riordan JR. CFTR function and prospects for therapy. *Annu Rev Biochem* **77**:701–726, 2008.

Schulenburg A, Ulrich-Pur H, Thurnher D, Erovic B, Florian S, Sperr WR, Kalhs P, Marian B, Wrba F, Zielinski CC and Valent P. Neoplastic stem cells: a novel therapeutic target in clinical oncology. *Cancer* **107**:2512–2520, 2006.

Schwiebert EM, Benos DJ, Egan ME, Stutts MJ and GugginoWB. CFTR is a conductance regulator as well as a chloride channel. *Physiol Rev* **79**:S145–S166, 1999.

Sebens S, Arlt A and Schafer H. NF-kappaB as a molecular target in the therapy of pancreatic carcinoma. *Recent Results Cancer Res* **177**:151–164, 2008.

Sekimoto M, Takada T, Kawarada Y, Hirata K, Mayumi T, Yoshida M, Hirota M, Kimura Y, Takeda K, Isaji S, Koizumi M, Otsuki M and Matsuno S. Japan guidelines for the management of acute pancreatitis: epidemiology, etiology, natural history, and outcome predictors in acute pancreatitis. *J Hepatobiliary Pancreat Surg* **13**:10–24, 2006.

Shapiro A, Ricordi C, Hering B, Auchincloss H, Lindblad R, Robertson R, Secchi A, Brendel M, Berney T, Brennan D, Cagliero E, Alejandro R, Ryan E, DiMercurio B, Morel P, Polonsky K, Reems J, Bretzel R, Bertuzzi F, Froud T, Kandaswamy R, Sutherland D, Eisenbarth G, Segal M, Preiksaitis J, Korbutt G, Barton F, Viviano L, Seyfert-Margolis V, Bluestone J and Lakey J. International trial of the Edmonton protocol for islet transplantation. *N Engl J Med* **355**:1318–1330, 2006.

Singh VP, Saluja AK, Bhagat L, van Acker GJ, Song AM, Soltoff SP, Cantley LC and Steer ML. Phosphatidylinositol 3-kinase-dependent activation of trypsinogen modulates the severity of acute pancreatitis. *J Clin Invest* **108**:1387–95, 2001.

Song AM, Bhagat L, Singh VP, Van Acker GG, Steer ML and Saluja AK. Inhibition of cyclooxygenase-2 ameliorates the severity of pancreatitis and associated lung injury. *Am J Physiol* **283**:G1166–1174, 2002.

Song SH and Hardisty CA. Early-onset Type 2 diabetes mellitus: an increasing phenomenon of elevated cardiovascular risk. *Expert Rev Cardiovasc Ther* **6**:315–322, 2008.

Stratta R and Alloway R. Pancreas transplantation for diabetes mellitus: a guide to recipient selection and optimum immunosuppression. *BioDrugs* **10**:347–357, 1998.

Stumvoll M, Goldstein BJ and van Haeften TW. Type 2 diabetes: principles of pathogenesis and therapy. *Lancet* **365**:1333–1346, 2005.

Tashiro M, Schäfer C, Yao H, Ernst SA and Williams JA. Arginine induced acute pancreatitis alters the actin cytoskeleton and increases heat shock protein expression in rat pancreatic acinar cells. *Gut* **49**:241–50, 2001.

Torgerson JS, Hauptman J, Boldrin MN and Sjostrom L. XENical in the prevention of diabetes in obese subjects (XENDOS) study: a randomized study of orlistat as an adjunct to lifestyle changes for the prevention of type 2 diabetes in obese patients. *Diabetes Care* **27**:155–161, 2004.

Tucker JA, Spock A, Spicer SS, Shelburne JD and Bradford W. Inspissation of pancreatic zymogen material in cystic fibrosis. *Ultrastruct Pathol* **27**:323–335, 2003.

Voltarelli J, Couri C, Stracieri A, Oliveira M, Moraes D, Pieroni F, Coutinho M, Malmegrim K, Foss-Freitas M, Simoes B, Foss M, Squiers E and Burt R. Autologous nonmyeloablative hematopoietic stem cell transplantation in newly diagnosed type 1 diabetes mellitus. *JAMA* **297**:1568–1576, 2007.

Voynow JA and Scanlin TF. Cystic fibrosis. In HB Panitch (ed), Pediatric Pulmonology. Elsevier Mosby, Philadelphia, PA, 2005.

Wang X, Wang SH and Gou H. The progress in the experimental studies of anticancer Chinese herbal compounds. *Chin J Exp Trad Med Form* **11**:71–73, 2005.

Wild S, Roglic G, Green A, Sicree R and King H. Global prevalence of diabetes: estimates for the year 2000 and projections for 2030. *Diabetes Care* **27**:1047–1053, 2004.

Zheng S and Chen A. Activation of PPARgamma is required for curcumin to induce apoptosis and to inhibit the expression of extracellular matrix genes in hepatic stellate cells in vitro. *Biochem J* **384**:149–157, 2004.

Zhou ZG, Yan WW, Chen YQ, Chen YD, Zheng XL and Peng XH. Effect of inducible cyclooxygenase expression on local microvessel blood flow in acute interstitial pancreatitis. *Asian J Surg* **27**:93–98, 2004.

Part II
The Renin-Angiotensin System (RAS)

Chapter 4
Circulating RAS

4.1 Definition of Circulating RAS

The classical renin-angiotensin system (RAS), along with aldosterone (also called RAAS), plays an important role in the regulation of blood pressure and fluid homeostasis (Peach, 1977; Reid et al., 1978). This classical or circulating RAS consists of several major components including the precursor angiotensinogen, two critical enzymes (renin and angiotensin-converting enzyme, ACE), their bioactive product, angiotensin II together with its receptors (AT_1 and AT_2 receptors), as well as aldosterone. Its regulatory actions on blood volume and systemic vascular resistance are mediated largely by stimulation of aldosterone released from the zona glomerulosa of the adrenal cortex to mediate renal sodium reabsorption and potassium excretion (and thus water retention), and by its potent vasoconstrictor effect on vascular smooth muscle, respectively (Lumbers, 1999; Matsusaka & Ichikawa, 1997). Notwithstanding the well-established existence of this classical RAS, the contemporary notion of this system has been fully revolutionized and revisited in recent decades, with particular emphasis on the growing candidates of new RAS components identified as well as with recognition of local RAS present in various tissue organs (Fyhrquist & Saijonmaa, 2008). This contemporary concept on the RAS will be critically reviewed and adequately discussed in the following sections and chapters of this book.

Generally speaking, our kidneys secrete renin in response to low blood pressure (hypotension), which stimulates the production of angiotensin I from angiotensinogen, which is synthesized from the liver. Circulating angiotensin I is, in turn, cleaved by the lung-derived ACE to produce angiotensin II, a physiologically active peptide of this system; being a potent vasoconstrictor, angiotensin II elicits blood vessels to constrict, thus resulting in increased blood pressure. On the other hand, angiotensin II, via the action on its receptors, notably the AT_1 receptor, located in the adrenal cortex, stimulates secretion of aldosterone, leading to Na^+ reabsorption, K^+ excretion, and water retention by the renal tubules, thereby increasing blood volume and systemic blood pressure. In addition, angiotensin II also stimulates anti-diuretic hormone (ADH) or vasopressin secretion from the posterior pituitary gland, which in turn increases water reabsorption by the renal collecting duct. The resultant increase

P.S. Leung, *The Renin-Angiotensin System: Current Research Progress in The Pancreas*, Advances in Experimental Medicine and Biology 690,
DOI 10.1007/978-90-481-9060-7_4, © Springer Science+Business Media B.V. 2010

in water and salt retention followed with concomitant increase in circulating blood volume negatively feedback onto the kidney, thereby inhibiting the release of plasma renin and thus circulating RAS activity (Giebisch & Windhager, 2003). In some physiological and pathophysiological conditions, there may be over-activation of the RAS, resulting in abnormally elevated blood pressure, i.e. hypertension and hypertension-associated disorders such as heart failure, kidney failure and diabetes. Of great interest of this context is the overactivity of pancreatic RAS induced by

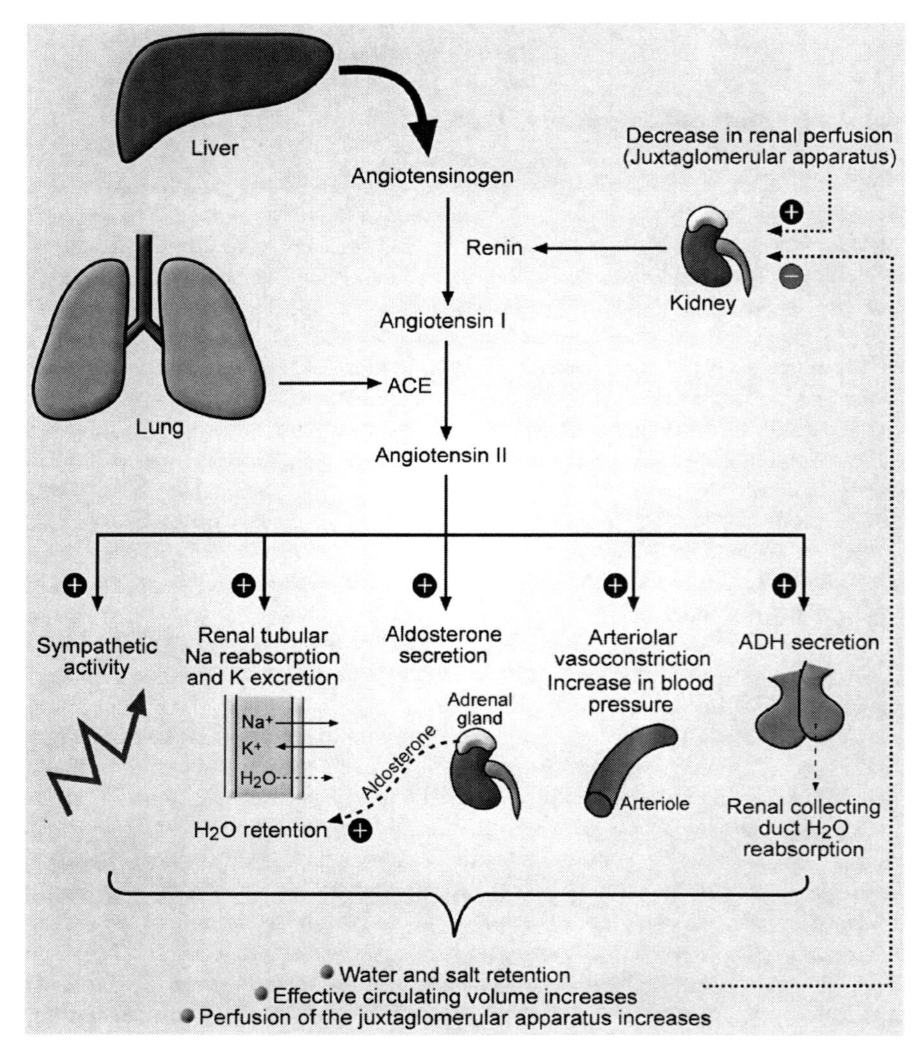

Fig. 4.1 A schematic diagram showing functioning of the RAAS in the regulation of blood pressure and fluid homeostasis (extracted from Wikipedia Online, 2010)

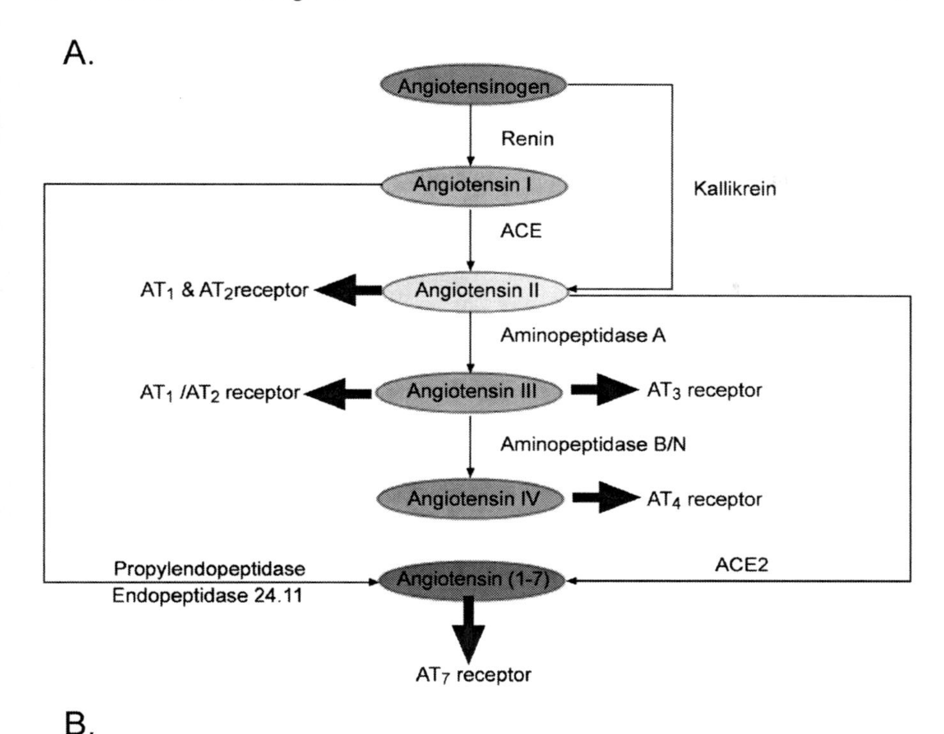

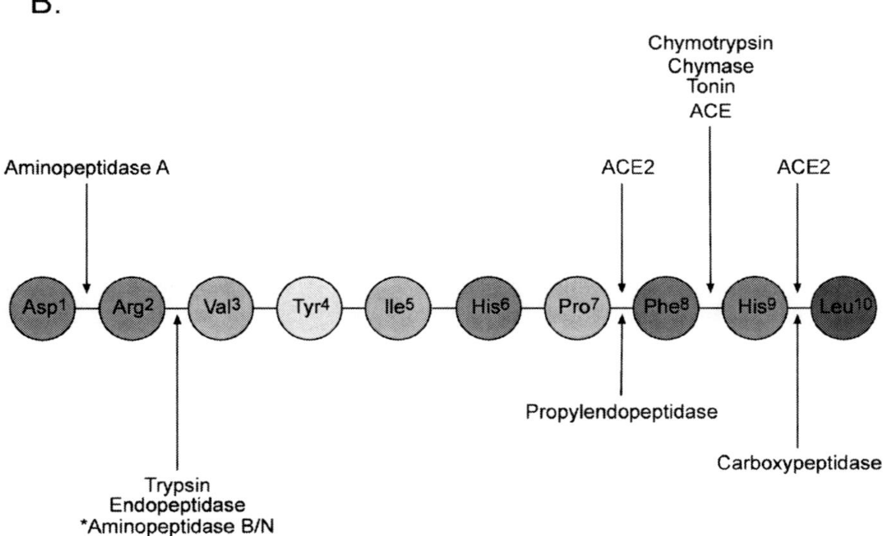

Fig. 4.2 Different angiotensin-processing enzymes for the generation of active angiotensin peptides (**a**) and some specific sites that are cleaved from the amino- and carboxy terminus as well as the interior of angiotensin I (**b**) (modified from Leung, 2006a)

some physiological and pathophysiological conditions, including hypoxia, pancreatitis, pancreatic inflammation, islet transplantation, hyperglycaemia and diabetes (T1DM and T2DM) and thus their clinical relevance (Leung, 2007), which will be discussed in the following chapters. A highly schematic depiction of the pathways by which the circulating RAAS controls blood pressure as well as electrolyte and water balance of our body is presented in Fig. 4.1.

In additional to the RAS components as discussed above, there are a raft of bioactive peptides, which are produced by the so-called alternate enzymes to renin and ACE, and they exhibit various biological actions via the mediation of their respective receptors. These alternate enzymes include, to name but a few, ACE2, chymase, kallikrein, aminopeptidase A and aminopeptidase B/N. Apart from angiotensin I (1–10) and angiotensin II (1–8), these alternate enzymes along with renin and ACE generate a host of such biologically active peptides as angiotensin III (2–8), angiotensin IV (3–8) and angiotensin (1–7). Angiotensin II and these bioactive peptides mediate their specific actions via respective trans-membrane receptors, such as AT_1 and AT_2 receptors, AT_3 receptor, AT_4 receptor and AT_7 receptor, located at target cells and tissues (Leung & Chappell, 2003; Leung, 2004). A critical appraisal of major RAS components and functions will be discussed in the following section below. The biosynthetic cascade for the RAS, its associated enzymes, bioactive peptide products, and their respective receptors are summarized in Fig. 4.2a.

4.2 Angiotensinogen

Angiotensinogen, also known as renin substrate, is the limiting factor in the enzymatic reaction of renin and an obligatory precursor of the circulating RAS. It is a serum glycoprotein with a molecular size ranging from 55 to 60 kDa that is synthesized and secreted mainly by the liver (Page et al., 1941). It is then released into the systemic circulation and the liver contains only small stores of angiotensinogen. Interestingly, this polypeptide is secreted constitutively and is not processed into secretory granules of the regulatory pathway and, consequently, most of its secretion is extracellular. In contrast, active renin is primarily transported into secretory granules for further processing in a regulatory fashion (see Section 4.3).

The angiotensinogen gene is present as a single copy in all species from human to rat and mouse (Gaillard et al., 1989; Ohkubo et al., 1983; Clouston et al., 1988). In the rat, the gene is 11.8 Kb long and is organized into five exons and four introns. The mRNA product consists of a coding sequence of 1,431 nucleotides flanked by 61 non-coding nucleotides at the 5′ end and 200–400 nucleotides at the 3′ end. Differences in this 3′ non-coding region arise from the presence of several polyadenylation sites which impairs heterogeneity in the size of the mRNA (Sernia, 1995). However, only a single protein product is formed, consisting of a 24 amino acid leader sequence and a 453 amino acid sequence for the mature angiotensinogen. The first 12 amino acids of angiotensinogen are the most important for its biological activity. The human angotensinogen gene contains five exons interrupted by four intervening sequences and the amino acid sequences, such as alpha 1-antitrypsin and antithrombin exhibit similarities when compared with that of rat in the intron-exon structure; the 5′-flanking sequence of the human angiotensinogen gene was examined for hormone regulatory elements, which may be associated with interaction with the hormone receptor complexes (Gaillard et al., 1989). On the other hand, protein separation of rat plasma via isoelectric focusing revealed two forms of angiotensinogen, namely A-1 and -2; the two angiotensinogen species have identical electrophoretic mobilities on analytical gel electrophoresis but differ in their apparent molecular weight, i.e. 60 and 56 kDa (Helgenfeldt & Hackenthal, 1982; Murakami et al., 1984). The heterogeneity of plasma angiotensinogen may be due to variable glycosylation but also represent a variation in amino acid sequence. In either case, the physiological significance of these two molecular species of angiotensinogen is still unknown. Nevertheless, previous studies have shown that the relatively low-molecular-weight form is the major component of circulating angiotensinogen while the larger form becomes more prevalent during the last trimester of pregnancy (Tewsbury & Dart, 1982).

Angiotensinogen is released into the circulatory system, where it becomes associated with the α_1-globulin, α_2-globulin or albumin fraction of plasma, depending on the species concerned (Plentl et al., 1943). Plasma angiotensinogen concentration ranges from 0.8 to 1.0 μM in human, 0.3 to 0.5 μM in rat, and 0.5 to 0.7 μM in rabbit (Reid et al., 1978). Plasma angiotensinogen levels are regulated by plasma corticosteroid, estrogen, and thyroid hormone, as well as angiotensin II. Specifically, its concentrations are increased after treatment with adrenocorticotrophic hormone, estrogen, and adrenal cortical hormones (Reid et al., 1973); in contrast, they are decreased after surgical procedures such as hypatectomy, adrenalectomy and hypophysectomy (Hasegawa et al., 1973; Reid et al., 1973). In additional to its hepatic source, angiotensinogen is widely expressed and localized in multiple tissues and organs, where it is subjected to regulation by various physiological and pathophysiological factors specific to its local functions in each tissue concerned. For example, angiotensinogen-generating tissues and organs include, but are not limited to, the brain (Greeland & Sernia, 2004), epididymis (Leung et al., 1999a, 2001), carotid body (Leung, 2006b; Lam & Leung 2003), intestine (Wong et al., 2007), and pancreas (Leung et al., 1999b; Regoli et al., 2003). The expression and functional correlates of angiotensinogen in some recently identified tissues at the local levels will be critically reviewed in Chapter 5.

4.3 Angiotensin-Generating Enzymes

The classical RAS has two critical enzymes, namely renin and ACE, which mediate sequential actions on the proteolytic cleavage of angiotensinogen into circulating angiotensin I and angiotensin II. Renin (EC 3.4.23.15) is an aspartyl protease and one of the key enzymes of the RAS. Unlike other aspartic proteases (i.e. pepsin or cathepsin D), renin is monospecific, acting selectively on angiotensinogen to produce inactive angiotensin I. It is synthesized as a zymogen pro-renin, which is subjected to proteolytic activation. The nucleotide sequence of renin has 10 exons in human, but only 9 in rodents. There is a high degree of sequence homology among renin isoforms (Hobart et al., 1984). Active renin cleaves its substrate angiotensinogen into angiotensin I. Pre(pro)renin and (pro)renin, the inactive precursors of renin, are found in circulating blood, amniotic fluid, and the kidney (Day & Luetscher, 1975). Acidification and low temperature can enhance angiotensin I-generating activity in plasma and amniotic fluid (Lumbers, 1971; Skinner et al., 1975). The afferent arteriolar juxtaglomerular cells of kidney act as the site of renin production for the RAS (Hackenthal et al., 1990). Synthesized pre(pro)renin is rapidly hydrolyzed into (pro)renin by signal protease. The (pro)renin, with a molecular weight 5 kDa greater than that of renin, is then converted to active renin and secreted in a regulated manner (Pratt et al., 1983). Expression of the renin gene is tissue-specific, occurring in mouse, but not rat, submandibular glands (Morris et al., 1980). Particularly important in the context of this book, renin mRNA and protein have been identified in several cells and tissues, notably, the vascular endothelium and islet β-cells of the pancreas (Leung et al., 1999b; Tahmasebi et al., 1999), which will be discussed in Chapters 8 and 10.

ACE (EC 3.4.15.1) is a dipeptidyl carboxypeptidase of the membrane-bound zinc ectoenzyme (also called peptidyl-dipeptidase A, peptidyl-dipeptidase I, kininase II, peptidase P, and carboxycathepsin), which was first identified as another key enzyme of the RAS. Its major functions are to convert plasma angiotensin I into physiologically active angiotensin II, and to degrade bradykinin by removal of a dipeptide from its C-terminus. There are also a number of dipeptide and tripeptide bonds that can be cleaved by ACE, such as those in metenkephalin, substance P, tachykinins, and prohormone convertase (Coates, 2003). As a counterpart of renin from the RAS, it is absolutely unambiguous that the classic role for ACE in human is a critical enzyme for the generation of active angiotensin II and thus appears to contribute mainly to maintenance of blood pressure and fluid homeostasis. In this context, two isoforms of ACE are also found to be expressed in mammals: a single-domain germinal isoform (gACE) required for male fertility, and a double-domain somatic isoform (sACE) which is critical for the RAS (Corvol et al., 1995). In short, the expression of gACE is closely associated with a functional implication in fertilization, apparently in reduced male fertility; sACE, in the capacity of a key enzyme of angiotensin II-generating enzyme, is critical for the regulation of blood pressure and electrolyte balance whilst, in the capacity of another processing enzyme, it is responsible for processing of several biologically active peptides such as tachykinins and prohormone convertase. Thus far, the clinical application of ACE has been mainly involved its pharmacological blockade (e.g. captopril and ramipril) in the control of hypertension, diabetic nephropathy and heart failure (Dell'Italia et al., 2002). In the pancreas, ACE expression, like renin, has also been identified in islet beta-cells and in the vascular endothelium of pancreatic islets (Carlsson et al., 1998), together with ACE activity and ACE mRNA (Ip et al., 2003) and its functional correlates will be again discussed in Chapters 8 and 10.

Apart from renin and ACE, a number of angiotensin-processing peptidases are capable to generate and metabolize biologically active angiotensin peptides and thus contribute to the so-called alternative pathways of the RAS that are different from the classical ones. These enzymes include, to name but a few, chymase, cathepsin G, chymotrypsin, trysin, tonin, kallikrein, ACE2 and other exopeptidases, as well as endopeptidases. These peptidases act directly and indirectly on angiotensin I and/or angiotensin II as well as the angiotensinogen precursor to generate a number of bioactive peptides with varying physiological activities, such as angiotensin (1–7), angiotengin (2–8)/angiotensin III, and angiotensin (3–8)/angiotensin IV, in additional to angiotensin (1–8)/angiotensin II (Fig. 4.2a). For example, angiotensin III is produced from angiotensin II by aminopeptidase A and exerts its action via AT_1 receptor, AT_2 receptor or AT_3 receptor; angiotensin IV is generated from angiotensin III by aminopeptidase B/N which exerts its action via AT_4 receptor (also called insulin-regulated membrane aminopeptidase, IRAP); angiotensin (1–7) heptapeptide is generated either from angiotensin I by endopeptidases (e.g. propylendopeptidase or endopeptidase 24.11) or from angiotensin II by ACE2 which exerts its action via AT_7 receptor (also called Mas receptor) (Campbell, 2003). Of particular interest in this context is the discovery of a novel peptidase termed ACE2, the first human homologue of ACE. Similar to ACE, ACE2 acts as a carboxypeptidase; however, ACE2 hydrolyzes a single residue in angiotensin II (Pro^7-Phe^8) or angiotensin I (His^9-Leu^{10}) to generate angiotensin (1–7) and angiotensin (1–9), respectively (Rice et al., 2004). ACE2 also cleaves some peptides, such as dynorphin, apelin and bradykinin. It has been previously thought that angiotensin (1–7) and angiotensin (1–9) are devoid of biological activities. Recently, a physiological role for ACE2-angiotensin (1–7)-Mas receptor axis has been implicated to have important roles in hypertension, heart function, and diabetes and, perhaps more importantly, as a receptor of the severe acute respiratory syndrome coronarvirus (Warner et al., 2004). The peptide linkages that are cleaved by different angiotensin-processing peptidases, with particular emphasis on ACE2, are depicted in Fig. 4.2b. Of note, the peptidase kallikrein, first isolated from dog and rat pancreata (Hojima et al., 1977), was found to be capable of generating angiotensin II directly from its precursor angiotensinogen (Arakawa, 1996). Indeed, several serine proteases capable of forming angiotensin II from angiotensin I and/or angiotensinogen, such as tonin, cathepsin G, and chyamase, have since been identified in the pancreas (Sasaguri et al., 1999; see also review by Fyhrquist & Saijonmaa, 2008).

4.4 Angiotensin and (Pro)renin Receptors

Most, if not all, of the RAS's major actions are mediated by physiologically active peptide angiotensin II via its two receptor subtypes, namely AT_1 and AT_2 receptors (De Gasparo et al., 2000). Both angiotensin II receptor subtypes are 7-transmembrane-spanning G-protein-coupled receptors. AT_1 and AT_2 receptors are comprised of 359 and 363 amino acids, respectively, and they share ~30% sequence similarity (Speth et al., 1995). Apart from their well-established role in regulation of blood pressure and fluid homeostasis, AT_1 and AT_2 receptors have recently been proposed to participate in novel, cell specific functions in tissue organs such as the pancreas and liver (Leung, 2004). These functions include stimulation and inhibition of cell proliferation, induction of apoptosis, generation of reactive oxygen species, regulation of hormone secretion, and proinflammatory and profibrogenic actions (Leung & Chappell, 2003). Of note, the actions of AT_1 receptor have antagonistic effects on those of AT_2 receptors. As discussed in subsequent chapters, it is indeed very clear that the expression of AT_1 receptors in the local RAS are generally regulated in opposite direction and the effects of AT_1 receptors are shown to counteract the effects of AT_2 receptor; this observation can be exemplified by exposure of the isolated mouse pancreatic islets to high glucose concentrations where AT_1 receptor expression is increased while reducing AT_2 receptor expression as well as function (Chu et al., 2010).

On the other hand, proteolytic fragments of angiotensin II also have biological activity via these and other receptors (Thomas & Mendelsohn, 2003). In this regard, angiotensin II can be metabolized into angiotensin III, which can act on AT_1 and AT_2 receptors, as well as an angiotensin III-specific receptor, AT_3 receptor (Chaki & Inagami, 1992). Angiotensin III has been proposed to be involved in chemokine production and cell growth regulation, via mediation of MCP-1 and NFκB and AP-1 expression (Ruiz-Ortega et al., 2000). It also plays a role in controlling blood pressure, thus serving as a putative target for treatment of hypertension (Reaux-Le Goazigo et al., 2005). Notwithstanding the existence of these supporting data, angiotensin III's roles remain however largely undefined. On the other hand, angiotensin III can be further metabolized into a hexapeptide called angiotensin IV, a bioactive ligand of the AT_4 receptor. The AT_4 receptor has a wide distribution in a range of tissues, including the brain (Chai et al., 2000). Interestingly, AT_4 receptor was recently identified to be the transmembrane enzyme, called IRAP, which is predominantly found in GLUT4 vesicles in insulin-responsive cells. Although the role of AT_4 receptor/IRAP has yet to be determined, it has been suggested to mediate memory and glucose uptake. The former might be attributed to IRAP prolonging the action of endogenous neuropeptides, while the latter could be due to modulation of GLUT4 trafficking (Chai et al., 2004).

Finally, a high affinity binding site for angiotensin (1–7) has been reported (Tallant et al., 1997), which is called AT_7 receptor or Mas receptor. Using a specific analogue for angiotensin (1–7), it has been possible to selectively block the binding site for angiotensin (1–7), but not for ACE. Several studies support the concept that angiotensin (1–7) induces vasodilation via activation of AT_7 receptors (Tom et al., 2003). However, solid evidence for the existence of AT_7/mas receptors in humans seems to be lacking. Interestingly, it is shown that ACE2-angiotensin (1–7)-mas receptor axis appears to represent a counter-regulatory arm of the RAS capable of protecting damaged tissues from the deleterious effects of angiotensin II which has been recently substantiated by the local hepatic RAS and fibrosis (Lubel et al., 2009). Despite of considerable evidence supporting the beneficial effects of

ACE2-angiotensin (1–7)-mas receptor axis in renal, cardiovascular and liver disease, the effect of this axis in the pancreas has yet to be investigated. In this context, it is quite intriguing that a "cross-talk" among the AT_2 receptor, bradykinin type 2 receptor (BK_2 receptor), and AT_7/mas receptor may exist in the RAS (Leung & Chappell, 2003). Some of the proposed functions of the angiotensin receptors (AT_1, AT_2, AT_3, AT_4 and AT_7) and their sites of potential cross-talk in the RAS are illustrated in Fig. 4.3.

Recent investigations have shown that (pro)renin may be functional in angiotensin independent pathways (Danser et al., 2007). This possibility was suggested by the discovery of a (pro)renin receptor, which can bind both renin and (pro)renin; this renin/(pro)renin receptor was recently cloned, characterized and shown to consist of a 350-amino acid protein with a single transmembrane domain (Nguyen et al., 2002). The binding of (pro)renin to the (pro)renin receptor, which occurs with greater affinity than renin binding of this receptor, renders (pro)renin catalytically active without proteolytic removal of its prosegment (Danser et al., 2007). On the other hand, (pro)renin may also exert as an agonist for this receptor, directly stimulating signalling pathways that are independent of angiotensin II (Saris et al., 2006). It has been reported that renin/(pro)renin triggers the activation of mitogen-activated protein kinase (MAPK) in mesangial cells, particularly p42/p44

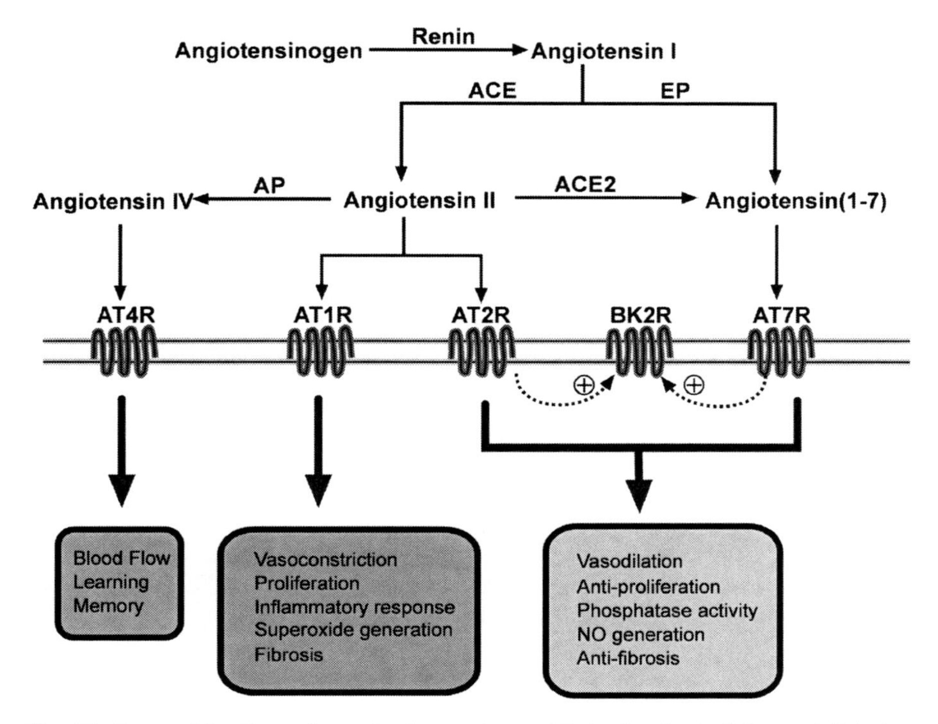

Fig. 4.3 Proposed functions of angiotensin receptors and their site of potential cross-talk in the RAS (modified from Leung & Chappell, 2003)

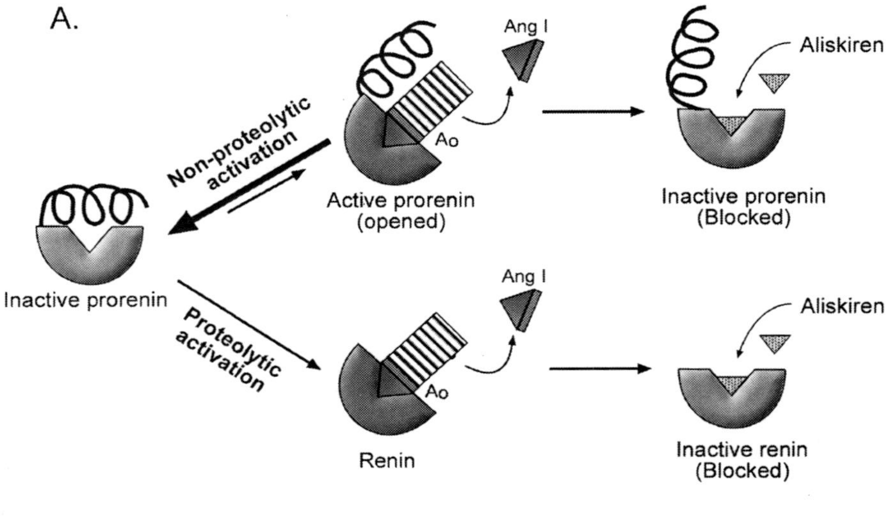

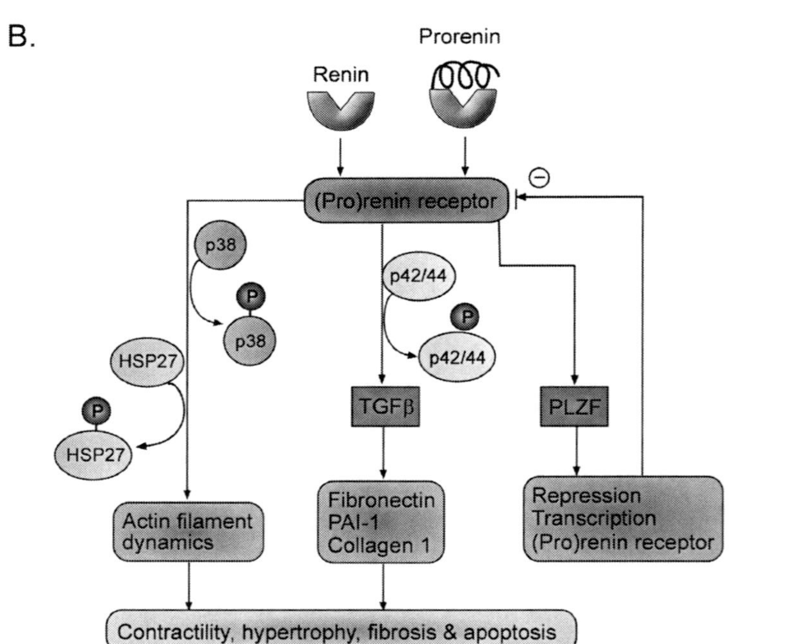

Fig. 4.4 A schematic diagram showing the proteolytic and non-proteolytic activation of prorenin (**a**) and putative actions of (pro)renin (**b**) (redrawn from Danser et al., 2007)

(Nguyen et al., 2002), and transforms growth factor-beta 1, thus leading to stimulation of plasminogen activator inhibitor, fibronectin, and collagen (Huang et al., 2006). These effects exist in the presence of renin inhibitors, ACE inhibitors and/or

AT_1 receptor blockers, suggesting the existence of receptor-mediated angiotensin II-induced independent mechanisms. Interestingly, over-expression of the human (pro)renin receptor leads to increased blood pressure, aldosterone levels and cyclooxygenase-2 expression in rat renal cortex; these changes do not associate with parallel changes in plasma renin activity or tissue angiotensin II content, thus further indicating that the (pro)renin receptor may directly stimulate signalling pathways in an angiotensin-independent manner (Danser et al., 2007). In view of these findings, it has been proposed that blockade of the renin/(pro)renin receptors (such as aliskiren) provides a new strategy for the treatment of renal and cardiovascular disease (Nguyen, 2007). Nevertheless, the potential role of renin/(pro)renin receptor in the physiology and pathophysiology of the pancreas still remains to be determined. The proteolytic and non-proteolytic activation of (pro)renin and its potential biological effects are summarized in Fig. 4.4.

References

Arakawa K. Serine protease angiotensin II systems. *J Hyperten* **14**:S3–S7, 1996.

Campbell DJ. The renin-angiotensin and the kallikrein-kinin systems. *Int J Biochem Cell Biol* **35**:784–791, 2003.

Carlsson PO, Berne C and Jansson L. Angiotensin II and the endocrine pancreas: effects on islet blood flow and insulin secretion in rats. *Diabetologia* **41**:127–133, 1998.

Chai SY, Fernando R, Peck G, Ye SY, Mendelsohn FA, Jenkins TA and Albiston AL. The angiotensin IV/AT4 receptor. *Cell Mol Life Sci* **61**:2728–2737, 2004.

Chai SY, Bastias MA, Clune EF, Matsacos DJ, Mustafa T, Lee JH, McDowall SG, Paxinos G, Mendelsohn FA and Albiston AL. Distribution of angiotensin IV binding sites (AT4 receptor) in the human forebrain, midbrain and pons as visualised by in vitro receptor autoradiography. *Chem Neuroanat* **20**:339–348, 2000.

Chaki S and Inagami T. A newly found angiotensin II receptor subtype mediates cyclic GMP formation in differentiated Neuro-2A cells. *Eur J Pharmacol* **225**:355–356, 1992.

Chu KY, Cheng Q, Chen C, Au LS, Seto SW, Tuo Y, Motin L, Kwan YW and Leung PS. Angiotensin II exerts glucose-dependent effects on K_V currents in mouse pancreatic beta cells via angiotensin II type 2 receptor. *Am J Physiol* **298**:C313–C323, 2010.

Clouston WH, Evans BA, Haralambidis J and Richard RI. Molecular cloning of the mouse angiotensinogen gene. *Genomics* **2**:240–248, 1988.

Coates D. The angiotensin converting enzyme (ACE). *Int J Biochem Cell Biol* **35**:769–773, 2003.

Corvol P, Williams TA and Soubrier F. Peptidyl dipeptidase an angiotensin-I converting enzyme. In AJ Barrett (ed), Proteolytic enzymes: aspartic and metallo peptidases, Academic Press, San Diego, CA, pp. 283–305, 1995.

Danser AHJ, Batenburg WW and van Esch JHM. Prorenin and the (pro)renin receptor. *Nephrol Dial Transplant* **22**:1288–1292, 2007.

Day RP and Luetscher JA. Occurrence of big renin in human plasma, amniotic fluid and kidney extracts. *J Clin Endocrinol Metab* **40**:1078–1084, 1975.

De Gasparo M, Catt KJ, Inagami T, Wright JW and Unger TH. The angiotensin II receptors. *Pharmacol Rev* **52**:415–472, 2000.

Dell'Italia LJ, Rocic P and Lucchesi PA. Use of angiotensin-converting enzyme inhibitors in patients with diabetes and coronary artery disease. *Curr Prob Cardiol* **27**:6–36, 2002.

Fyhrquist F and Saijonmaa O. Renin-angiotensin system revisited. *J Int Med* **264**:224–236, 2008.

Gaillard I, Clauser E and Corvol P. Structure of the human angiotensinogen gene. *DNA* **8**:87–99, 1989.

Giebisch H and Windhager E. Integration of salt and water balance. In WF Boron and EL Boulpaep (eds), Medical physiology, Saunders, New York, NY, pp. 861–876, 2003.
Greeland KJ and Sernia C. Oestrogenic regulation of brain angiotensinogen. *J Neuroendocrinol* **16**:508–515, 2004.
Hackenthal E, Paul M, Ganten D and Taugner R. Morphology, physiology, and molecular biology of renin secretion. *Physiol Rev* **70**:1067–1116, 1990.
Hasegawa HA, Nasjletti A, Rice K and Masson GM. Role of pituitary and adrenals in the regulation of plasma angiotensinogen. *Am J Physiol* **225**:1–6, 1973.
Helgenfeldt U and Hackenthal E. Separation and characterization of two different species of rat angiotensinogen. *Biochem Biophys Act* **708**:335–342, 1982.
Hobart PM, Fogliano M, O'connor BA, Schaefer IM and Chirgwin JM. Human renin gene: structure and sequence analysis. *Proc Natl Acad Sci USA* **81**:5026–5030, 1984.
Hojima Y, Yamashita N, Ochi N, Moriwaki C and Moriya H. Isolation and properties of dog and rat pancreatic kallikreins. *J Biochem* **81**:599–610, 1977.
Huang Y, Wongamorntham S, Kasting J. Renin increases mesangial cell transforming factor-beta 1 and matrix proteins through receptor-mediated angiotensin II-independent mechanisms. *Kidney Int* **69**:105–113, 2006.
Ip SP, Kwan PC, Williams CH, Pang S, Hooper NM and Leung PS. Changes of angiotensin-converting enzyme activity in the pancreas of chronic hypoxia and acute pancreatitis. *Int J Biochem Cell Biol* **35**:944–954, 2003.
Lam SY and Leung PS. Chronic hypoxia activates a local angiotensin-generating system in the carotid body. *Mol Cell Endocrinol* **203**:147–153, 2003.
Leung PS. The peptide hormone angiotensin II: its new functions in tissues and organs. *Curr Protein Pept Sci* **5**:267–273, 2004.
Leung PS. Novel roles of a local angiotensin-generating system in the carotid body. *J Physiol* **575**:4, 2006a.
Leung PS. Importance of the local renin-angiotensin system in pancreatic disease. In: U Lendeckel and NM Hooper (eds), Protease in gastrointestinal tissues, vol 5, pp. 131–152, Springer, Dordrecht, 2006b.
Leung PS. The physiology of a local renin-angiotensin system in the pancreas. *J Physiol* **580**: 31–37, 2007.
Leung PS and Chappell MC. A local pancreatic renin-angiotensin system: endocrine and exocrine roles. *Int J Biochem Cell Biol* **35**:838–846, 2003.
Leung PS, Fung ML and Sernia C. Chronic hypoxia induced down-regulation of angiotensinogen in rat epididymis. *Regul Pept* **96**:143–149, 2001.
Leung PS, Wong TP and Sernia C. Angiotensinogen expression by rat epididymis: evidence for an intrinsic, angiotensin-generating system. *Mol Cell Endocrinol* **155**:115–122, 1999a.
Leung PS, Chan WP, Wong TP and Sernia C. Expression and localization of the renin-angiotensin system in the rat pancreas. *J Endocrinol* **160**:13–19, 1999b.
Lubel JS, Herath CB, Tchongue J, Grace J, Jia Z, Spencer K, Casley D, Crowley P, Sievert W, Burrell LM and Angus PW. Angioensin-(1–7), an alternative metabolite of the renin-angiotensin system, is up-regulated in human liver disease and has anti-fibrotic activity in the bile-duct-ligated rat. *Clin Sci* **117**:375–386, 2009.
Lumbers ER. Angiotensin and aldosterone. *Regul Pept* **80**:91–100, 1999.
Lumbers ER. Activation of renin in human amniotic fluid by low pH. *Enzymologia* **40**:329–336, 1971.
Matsusaka T and Ichikawa I. Biological functions of angiotensin and its receptors. *Ann Rev Physiol* **59**:395–412, 1997.
Morris BJ, De Zwart RT and Young JA. Renin in mouse but not in rat submandibular glands. *Experientia* **36**:1333–1334, 1980.
Murakami E, Eggena P, Barret JD and Sambhi MP. Heterogeneity of renin substrate released from hepatocytes and in brain extracts. *Life Sci* **34**:385–392, 1984.

Nguyen G, Delarue F, Burckle C, Bouzhir L, Giller T and Sraer JD. Pivotal role of the renin/prorenin receptor in angiotensin II production and cellular response to renin. *J Clin Invest* **109**:1417–1427, 2002.

Nguyen G. The (pro)renin receptor: pathophysiological roles in cardiovascular and renal pathology. *Curr Opin Nephrol Hypertens* **16**:129–133, 2007.

Ohkubo H, Kageyama R, Ujihara M, Hirose T, Inayama S and Nakanjshi S. Cloning and sequence analysis of cDNA for rat angiotensinogen. *Proc Natl Acad Sci USA* **80**:3069–3073, 1983.

Page IH, McSwain B, Knapp GM and Andrus WD. The origin of renin activator. *Am J Physiol* **135**:214–222, 1941.

Peach M. Renin-angiotensin system: biochemistry and mechanism of action. *Physiol Rev* **57**: 313–370, 1997.

Plentl AA, Page IH and Davis WW. The nature of the renin activator. *J Biol Chem* **147**:143–153, 1943.

Pratt RE, Ouellette AJ and Dzau VJ. Biosynthesis of the renin: multiplicity of active and intermediate forms. *Proc Natl Acad Sci USA* **80**:6909–6813, 1983.

Reaux-Le Goazigo A, Iturrioz X, Fassot C, Claperon C, Roques BP and Llorens-Cortes C. Role of angiotensin III in hypertension. *Curr Hypertens Rep* **7**:128–134, 2005.

Regoli M, Bendayan M, Fonzi L, Sernia C and Bertelli E. Angiotensinogen localization and secretion in the rat pancreas. *J Endocrinol* **179**:81–89, 2003.

Reid IA, Morris BJ and Ganong WF. The renin-angiotensin system. *Annu Rev Physiol* **40**:377–410, 1978.

Reid IA, Tu WH and Ostsuka K. Studies concerning the regulation and importance of plasma angiotensinogen concentration in the dog. *Endocrinology* **93**:107–114, 1973.

Rice GI, Thomas DA, Grant PJ, Turner AJ and Hooper NM. Evaluation of angiotensin-converting enzyme (ACE), its homologue ACE2 and neprilysin in angiotensin peptide metabolism. *Biochem J* **383**:45–51, 2004.

Ruiz-Ortega M, Lorenzo O and Egido J. Angiotensin III increases MCP-1 and activates NF-kappa B and AP-1 in cultured mesangial and mononuclear cells. *Kid Int* **57**:2285–2298, 2000.

Saris JJ, 't Hoen PA, Garrelds IM, Dekkers DH, den Dunnen JT, Lamers JM and Jan Danser AH. Prorenin induces intracellular signaling in cardiomyocytes independently of angiotensin II. *Hypertension* **48**:564–571, 2006.

Sasaguri M, Noda K, Tsuji E, Koga M, Kinoshita A, Ideishi M, Ogata S and Arakawa K. Structure of a kallikrein-like enzyme and its tissue localization in the dog. *Immunopharmacology* **44**: 15–19, 1999.

Sernia C. Location and secretion of brain angiotensinogen. *Regul Pept* **57**:1–18, 1995.

Skinner SL, Cran EJ, Gibson R, Taylor R, Walters WA and Catt KJ. Angiotensins I and II, active and inactive renin, renin substrate, renin activity, and angiotensinase in human liquor amnii and plasma. *Am J Obstet Gynecol* **121**:626–630, 1975.

Speth RC, Thompson SM and Johns SJ. Angiotensin II receptors: structural and functionalconsiderations. In AK Mukhopadhyay and MK Raizada (eds), Tissue renin-angiotensin systems, Plenum Press, New York, NY, pp. 169–192, 1995.

Tahmasebi M, Puddefoot JR, Inwang ER and Vinsion GP. The tissue renin-angiotensin systemin human pancreas. *J Endocrinol* **161**:317–322, 1999.

Tallant EA, Lu X, Weiss RB, Chappell MC and Ferrario CM. Bovine aortic endothelial cells contain an angiotensin (1–7) receptor. *Hypertension* **29**:388–393, 1997.

Tewsbury DA and Dart RA. High molecular weight angiotensinogen levels in hypertensive pregnant women. *Hypertension* **4**:729–734, 1982.

Thomas WG and Mendelsohn FAO. Angiotensin receptors: form and function and distribution. *Int J Biochem Cell Biol* **35**:774–779, 2003.

Tom B, Dendorfer A and Danser AHJ. Bradykinin, angiotensin-(1–7), and ACE inhibitors: how they interact. *Int J Biochem Cell Biol* **35**:792–801, 2003.

Warner FJ, Smith AI, Hooper NM and Turner AJ. Angiotensin-converting enzyme-2: a molecular and cellular perspective. *Cell Mol Life Sci* **61**:2704–2713, 2004.

Wikipedia online. Renin-Angiotensin System. http://en.wikipedia.org/wiki/Renin-angiotensin_system. 2010

Wong TP, Debnam ES and Leung PS. Involvement of an enterocyte renin-angiotensin system in the local control of SGLT1-dependent glucose uptake across the rat small intestinal brush border membrane. *J Physiol* **584**:613–623, 2007.

Chapter 5
Local RAS

5.1 Definition of the Local RAS

The concept of a circulating RAS is well established and known to play an endocrine role in the regulation of fluid homeostasis (see Section 4.1, Chapter 4). However, it is more appropriate to view the RAS in the contemporary notion as an "angiotensin-generating system", which consists of angiotensinogen, angiotensin-generating enzymes, and angiotensins, as well as their receptors. Some RASs can be termed as "complete", having renin and ACE involved in the biosynthesis of angiotensin II peptide, i.e. in a renin and/or ACE-dependent manner which is exemplified in the circulating RAS. On the other hand, some RAS can be termed as "partial", having alternate enzymes to renin and ACE, such as chymase and ACE2 (see Section 4.3, Chapter 4) available for the generation of angiotensin II and other bioactive angiotensin peptides in the biosynthetic cascade, i.e. in a renin and/or ACE-independent manner. Complete vs. partial RASs can be exemplified in the so-called intrinsic angiotensin-generating system or local RAS; for example, a local and functional RAS with renin and ACE-dependent but a renin-independent pathway have been indentified in the pancreas (Leung, 2007) and carotid body (Lam & Leung, 2002), respectively. In the past two decades, local RASs have gained increasing recognition especially with regards to their clinical importance. Distinct from the circulating RAS, these functional local RASs exist in such diverse tissues and organs as the pancreas, liver, intestine, heart, kidney, vasculature, carotid body, and adipose, as well as the nervous, reproductive, and digestive systems (Paul et al., 2006). Taken into previous findings from our laboratory and others together, Table 5.1 is a summary of some recently identified local RASs in various levels of tissues and organs.

Local RASs can operate in an autocrine, paracrine and/or intracrine manner, and exhibit multiple physiological effects at the cellular and tissue levels in addition to and distinct from those of the circulating RAS (Lavoie & Sigmund, 2003). In addition to hemodynamic actions, local RASs have multiple, novel functions including regulation of cell growth, differentiation, proliferation and apoptosis, reactive oxygen species (ROS) generation, tissue inflammation and fibrosis, and hormonal secretion (Leung, 2004). Such a diversity of roles makes local RASs attractive

P.S. Leung, *The Renin-Angiotensin System: Current Research Progress in The Pancreas*, Advances in Experimental Medicine and Biology 690,
DOI 10.1007/978-90-481-9060-7_5,

Table 5.1 Some examples of recently identified intrinsic angiotensin-generating systems in different tissues and organs

Tissues and organs	Components of local angiotensin-generating system
Heart	Renin, ACE, Chymase, Angiotensinogen, Ang I, Ang II, AT_1 & AT_2 receptors
Vasculature	Renin, ACE, ACE2, Angiotensinogen, Ang I, Ang II, AT_1 & AT_2 receptors
Brain	Renin, ACE, Tonin, Cathepsin, Chymase, Angiotensinogen, Ang I, Ang II, Ang IV, Ang (1–7), AT_1, AT_2 & AT_4 receptors
Ovary	Renin, ACE, Angiotensinogen, Ang I, Ang II, AT_1 & AT_2 receptors
Uterus	Renin, ACE, Angiotensinogen, Ang I, Ang II, AT_1 & AT_2 receptors
Testis	Renin, ACE, Angiotensinogen, Ang I, Ang II & AT_1 receptors
Epididymis	Renin, ACE, Angiotensinogen, Ang I, Ang II, AT_1 & AT_2 receptors
Liver	Renin, ACE, Angiotensinogen, AT_1 & AT_2 receptors
Pancreas	Renin, ACE, Angiotensinogen, Ang I, Ang II, AT_1 & AT_2 receptors
Intestine	Renin, ACE, ACE2, Angiotensinogen, Ang II, AT_1 & AT_2 receptors
Adipose	Renin, ACE, Angiotensinogen, Ang I, Ang II & AT_1 receptors
Carotid body	ACE, Angiotensinogen, Ang II, AT_1, AT_2 & AT_4 receptors

therapeutic targets in diverse disease states. In the following sections, three of these functional local RASs, which have been characterized and investigated in my laboratory, will be presented for discussion and critical appraisal of their physiological roles and clinical relevances.

5.2 Local RAS in Carotid Body

The carotid body is a highly vascularised network of chemoreceptor cells, located at the bifurcation of the internal and external carotid arteries from the common carotid artery. Carotid body chemoreceptors, called glomus (or type 1, catechol-secreting) cells, release dopamine in response to rapid changes in arterial oxygen pressure (PO_2), carbon dioxide pressure (PCO_2), and pH, thus increasing carotid body sinus nerve activity and provoking centrally-mediated cardiopulmonary responses (Gonzalez et al., 1994). In this context, the carotid body is critical for physiological responses to acute, chronic and intermittent hypoxia. Hypoxia activation of the RAS might be associated with chronic cardiopulmonary diseases and sleep apnea. Of great interest in this context is emerging evidence for the existence of a local carotid angiotensin-generating system, which is upregulated by chronic hypoxia to alter carotid chemoreceptor sensitivity.

5.2.1 Expression and Function of Carotid Body RAS

As mentioned, newly discovered functions of intrinsic angiotensin-generating systems have been identified in many tissues and organs and are of particular physiological and pathophysiological interest. Such systems have recently been proposed to exist in the carotid body, where key RAS components including angiotensinogen, ACE, AT_1 and AT_2 receptors as well as AT_4 receptors (Table 5.1) are localized and expressed in the absence of renin, thus suggesting the existence of a renin-independent biosynthetic pathway for angiotensin II in the carotid body (Lam & Leung, 2002). The local RAS in carotid body plays a key role in modulating the carotid body response to hypoxia. Angiotensin II itself increases carotid body efferent activity, presumably via activation of the AT_1 receptor (Allen, 1998). Meanwhile, chronic hypoxia upregulates several key RAS components in the carotid body, including time-course dependent ACE activity (Lam & Leung 2003; Lam et al., 2004). Specifically, AT_1 receptor expression on glomus cells is enhanced, consistent with increased transcription and translation of the AT_1 receptor (Leung et al., 2000). Chronic hypoxia thus enhances AT_1 receptor-mediated efferent activity (Leung et al., 2000). In addition, AT_1 receptor activation increases intracellular calcium levels in dissociated glomus cells, an effect which is enhanced threefold by chronic hypoxaemia (Fung et al., 2001). Interestingly, postnatal hypoxemia is associated with an increased sensitivity of peripheral chemoreceptors in response to angiotensin II; in this condition, there is an upregulation of AT_{1a} receptor-mediated intracellular calcium activity, but a down-regulation of AT_{1b} receptor, of the chemoreceptors. These data may be important for adaptation of carotid body functions in the hypoxic ventilatory response and in electrolyte and water balance during perinatal and postnatal hypoxia (Fung et al., 2002).

High-affinity binding sites for angiotensin IV, known as AT_4 receptors (also called insulin-regulated membrane aminopeptidase, IRAP) were found to be localized and expressed in various tissues and organs (Albiston et al., 2001). As discussed in Chapter 4, angiotensin IV, a pentapeptide containing a 3–8 peptide fragment from an angiotensin II metabolite, is a biologically active peptide of the RAS which exerts its action via mediation of AT_4 receptor. Interestingly, AT_4 receptors are expressed and localized in the rat carotid body and, more importantly, play a prominent role following its upregulation in chronic hypoxia (Fung et al., 2007). This observation is consistent with the fact that the expression and function of local RAS components, notably AT_1 receptors and ACE in the carotid body, are upregulated by chronic hypoxia (Leung et al., 2000; Lam & Leung 2003; Lam et al., 2004). In addition, activation of AT_4 receptors can increase local production of angiotensin II in the carotid body, which in turn elevates angiotensin IV and thus activates an intracellular signalling cascade leading to increased intracellular calcium concentrations. It is plausible to speculate that the activation of AT_4 receptors in the carotid body may serve as an additional pathway enhancing the angiotensin II action in the activation of the chemoreflex in the physiological response to chronic hypoxia (Fung et al., 2007). Taken together, all of these available data suggest that chronic hypoxia is a major factor in increasing major RAS component expression and function in the carotid chemoreceptor, with likely changes in cardiopulmonary function. Thus the excitatory action of angiotensin II on the glomus cells may increase the chemosensitivity of the carotid body and may counteract the blunting effect of chronic hypoxia. Notwithstanding this involvement of AT_4 receptor-mediated carotid body function, details of the molecular and cellular mechanisms underlying the modulation of this receptor require extensive study.

5.2.2 Carotid Body RAS and Congestive Heart Failure

In a recent report using an animal model of congestive heart failure (CHF), a role for local glomus angiotensin II/AT_1 receptor signaling in increasing the sensitivity of Kv channels to hypoxia has been shown (Li & Schultz, 2006). In this study, high concentration of angiotensin II (>1 nM) directly inhibits Kv currents (I_k) while changes in Kv channel protein expression may contribute to the suppression of I_k and enhanced sensitivity of I_k to hypoxia in a CHF state. These findings have two significant implications in term of clinical setting: angiotensin II can enhance the oxygen sensitivity of Kv channels via the AT_1 receptor, and the cellular angiotensin

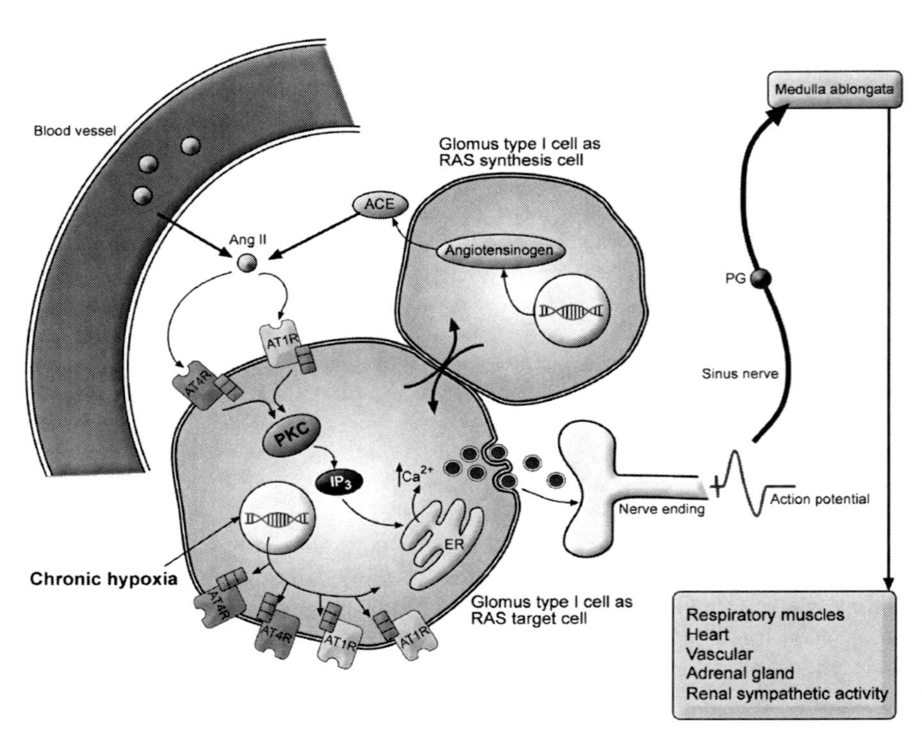

Fig. 5.1 Representation of the relationship of circulating RAS and carotid RAS involved in the regulation of cardiopulmonary function (modified from Leung et al., 2003)

II/AT_1 receptor pathway is functional in carotid glomus cells of CHF, but not normal animals. In spite of this solid evidence, the precise mechanism by which CHF upregulates the angiotensin II/AT_1 receptor and its function in the carotid body have yet to be resolved. In this regard, chronically impaired cardiac output resulting from CHF might be sufficient to render a prolonged deprivation of oxygen delivery to carotid glomus cells, a state akin to chronic hypoxia, thus leading to activation of its cellular RAS pathway. These data suggest that angiotensin II may have a paracrine and/or autocrine role in the modulation of enhanced carotid chemoreceptor sensitivity to chronic hypoxia characterized in CHF (Leung, 2006).

Taken together, the aforementioned findings suggest that chronic hypoxia is a major factor that increases major RAS component expression in the carotid body and activity of the carotid chemoreceptors, likely accompanied by changes in cardiopulmonary function. Thus the excitatory action of angiotensin II and its actions via AT_1 and AT_4 receptors on the glomus cells may increase the chemosensitivity of the carotid body while counteracting the blunting effect of chronic hypoxia. Notwithstanding the evidence of a functional angiotensin-generating system in the carotid body, details of the molecular and cellular mechanism(s) underlying the modulation require extensive further investigations. The potential interactions between circulating and local RAS-mediated effects of the carotid body in regulating cardiopulmonary function during normoxia and hypoxia are briefly summarized in Fig. 5.1.

5.3 Local RAS in Liver

The liver is a metabolically active, highly aerobic organ. It receives approximately 30% of the total blood flow and extracts 20% of the oxygen used by the human body (Hardikar & Suchy, 2003). Liver functions include, but are not limited to: (1) filtration and storage of blood; (2) metabolism of carbohydrates, proteins, fats, hormones, and foreign chemicals; (3) formation of bile; (4) storage of vitamins and iron; (5) formation of coagulation factors; (6) synthesis of hormone precursors, such as angiotensinogen; (7) excretion and degradation of hormones, drugs and toxins; and (8) exocrine bile secretion. The liver is composed mainly of hepatocytes, or liver cells, which represent about 60% of the total cell population. In addition to the hepatic cells, the venous sinusoids are lined by three other types of cells, namely the typical endothelial cells, large Kupffer cells, and hepatic stellate cells. Kupffer cells are a type of macrophage capable of phagocytizing bacteria and other foreign matter in the hepatic sinus blood (Alison, 1986). Hepatic stellate cells represent a small and versatile population cell type in the liver. Its major function of resting or quiescent hepatic cell is to store vitamin A whilst, during activation by various stimuli (e.g. injured hepatocytes, activation of kupffer cells and inflammatory cytokine and reactive oxygen species), hepatic stellate cells acquire a myofibroblast-like phenotype which plays a key role in the pathogenesis of liver fibrosis, including collagen deposition and abnormal extracellular matrix remodelling (Atzori et al., 2009).

5.3.1 RAS and Liver Function

Angiotensinogen is best known to be synthesized in the liver and it is the obligatory component of the circulating and local RAS as discussed in previous sections. Systemic and local RASs regulate liver function and liver disease. This role is particularly true for the local hepatic RAS, which remains largely ambiguous in the liver. In this regard, our recent findings have demonstrated the expression of several RAS components in the liver. Some, but not all, of the RAS components, including renin, ACE, and, in particular, the two AT_1 receptor subtypes (AT_{1a} and AT_{1b}), are expressed in Kupffer cells. The precursor angiotensinogen, the mandatory component for a locally generating-angiotensin system, however, is *not* present in Kupffer cells. While both renin and ACE are present, the AT_2 receptor appears not to exist in Kupffer cells (Table 5.2). These data lend support for the existence of an angiotensin-generating system in the liver and a potential role of angiotensin II in hepatic Kupffer cells (Leung et al., 2003). Interestingly, our observations further suggest that angiotensin II could upregulate the expression of TGF-β and fibronectin, and that a local hepatic RAS acting via AT_1 receptors located on Kupffer cells may play a role in the regulation of a fibrogenic action. Using a D-galactosamine-induced rat model of liver failure, AT_1 receptor blockade was demonstrated to prevent the progression of acute liver injury, as evidenced by an improvement in survival, reduction of liver enzymes, hepatic histopathology, and reduction in fibrogenic tissue-specific inhibitor of metalloproteinsase protein (Chan et al., 2007). These findings further indicate that hepatic AT_1 receptors may mediate liver inflammatory responses in liver injury. On the other hand, the RAS is also known to play an important role in chronic liver diseases, including liver cirrhosis and non-alcoholic steatohepatitis (NASH), the latter being the most frequent cause of chronic liver impairment (Lubel et al., 2008). In fact, the RAS is closely involved in liver fibrosis through activation of hepatic stellate cells, which mediates their fibrogenic actions by over-expression of profibrogenic factors in the liver. In an animal model of NASH, it has been shown that inhibition of the AT_1 receptor attenuates aspartate aminotransferase levels, activation of hepatic stellate cells, oxidative stress, expression of transforming growth factor-β1, expression of collagen genes, and liver fibrosis (Hirose et al., 2007).

Table 5.2 Comparison of the presence and absence of major RAS component expression in the liver and in the Kupffer cells

RAS component	Kupffer cells	Liver
Angiotensinogen	–	+++
AT_{1a} receptor	+++	++
AT_{1b} receptor	+	++
AT_2 receptor	–	++
Renin	++	++
ACE	+	+

+++, Very strong expression; ++, strong expression; +, weak expression; –, no expression.

In addition to activation of the AT_1 receptor, several RAS components are also upregulated in an experimental model of liver fibrosis, including angiotensin (1–7) and ACE2 (Herath et al., 2007). Indeed ACE2 expression is elevated in injured liver, in both humans and rats, which in turn regulates RAS activity in liver fibrosis (Paizis et al., 2005). In a rat model of cirrhosis with bile duct ligation, it has been recently demonstrated that angiotensin (1–7) is highly upregulated in human liver disease and has anti-fibrotic action, further supporting the notion that the ACE2-angiotensin (1–7)-mas receptor axis represents a potential target for anti-fibrotic therapy observed in liver cirrhosis (Lubel et al., 2009). In addition, administration of angiotensin II increases portal pressure in experimental and human cirrhosis, whereas AT_1 receptor antagonism abrogates the angiotensin II effect. These data suggest that angiotensin II may be involved in the pathogenesis of portal hypertension in cirrhosis (Rockey & Weisiger, 1996, Schneider et al., 1999). Angiotensin II has also been shown to exert its contractive action on the postsinusoidal venules (Arroyo et al., 1981), and the localization of AT_1 receptors in hepatic stellate cells gives new insight into the direction of RAS research with respect to regulation of liver fibrosis (Bataller et al., 2000). Recent studies have also shown that angiotensin II can induce contraction and proliferation of hepatic stellate cells via the AT_1 receptor. Taken together, these data indicate that angiotensin II plays a pivotal role in the development of liver fibrosis through the activation of hepatic stellate cells (Yoshiji et al., 2001). On the other hand, Kupffer cells have been known to get actively involved in fibrotic processes of the liver (Casini, 2000). Indeed, Kupffer cells constitute 90% of the sessile macrophages of the reticuloendothelial system and they are a major source of cytokines (Poli, 2000). Cytokines are believed to activate hepatic stellate cells, induce the proliferation and differentiation of stellate cells, and stimulate the production of extracellular matrix components of hepatic stellate cells, such as TGF-β and fibronectin. Despite of the above convergent evidence, the potential role of the RAS and its clinical relevance in liver disease are yet to be determined; this uncertainty is particularly true for the local hepatic RAS, which remains largely ambiguous and to be investigated.

5.3.2 Interaction Between Hepatic RAS and Vitamin D in T2DM

As just mentioned in the early beginning of Section 5.3, the liver is a vital organ containing several cell types, namely hepatocytes, stellate and Kupffer cells, with a central role in glucose metabolism and it is a key target for insulin action. As such, it is unsurprising that defects in liver cell function should increase the risks of type 2 diabetes mellitus (T2DM). Meanwhile, vitamin D is known best for its calcium and bone homeostasis but hypovitaminosis D, or vitamin D deficiency, also increases insulin resistance, impairs insulin secretion and is associated with, and predicts, all metabolic syndrome components, T2DM and also non-alcoholic fatty disease and fibrosis of the liver (NAFLD, a recognised part of the metabolic syndrome). In keeping with these, hypovitaminosis D is linked with enhanced risks of NAFLD (Targher et al., 2007) and hepatic fibrosis with cirrhosis developing in 10% of such patients (Harrison, 2006). Hypovitaminosis D is associated with increased risks of each component of metabolic syndrome: central obesity, glycemia, blood pressure, adverse lipid profiles, insulin resistance, T2DM, and cardiovascular disease (see review by Boucher, 1998). Mechanisms explaining vitamin D's effects on insulin sensitivity are not fully elucidated but, as for islet cell function, may include increasing intracellular ionised calcium or modulation of the many genes with vitamin D receptor response elements in their promoter regions (Wang et al., 2005). NAFLD increases hepatic insulin resistance, independent of general obesity in humans while vitamin D inhibits adipogenesis via vitamin D receptor (VDR)-mediated inhibition of PPARγ activity (Woo et al., 2008). Vitamin D is both 25-hydroxylated and activated in liver (Hollis, 1990) by local 1α-hydroxylase and there are fully functional VDRs in Kupffer, stellate and endothelial cells as well as in the hepatocytes (Gascon-Barre et al., 2003); thus vitamin D could be expected to exert direct effects on hepatic insulin signalling pathway genes.

Concurrently, several clinical studies have reported inverse relationships between circulating concentrations of activated vitamin D and blood pressure in patients with high-renin activity hypertension (Lind et al., 1995; Burgess et al., 1990). VDR-null mice have been shown to have a defect with increases in renin expression consistent with the suggestion that vitamin D is a negative regulator of the RAS (Li et al., 2002). A potential mechanism for this effect has subsequently been demonstrated, showing that the angiotensin II feedback suppression of renin expression was different in VDR-null mice compared with wild type mice (Kong & Li, 2003). In light of these background findings, it is plausible to suggest that vitamin D, via direct actions and by modulation of RAS activity, could exert effects on hepatic insulin sensitivity, NAFLD and metabolic syndrome such as T2DM. As a first step to test our hypothesis, we first examine the localization, expression and regulation of major RAS components in an animal model of obesity-induced db/db diabetic mouse. Protein levels of AT_1 receptor, ACE and renin in obese diabetic mice are significantly upregulated while AT_2 receptor is down-regulated when compared with their respective control livers, as evidenced by Western blot analysis (Fig. 5.2). By means of double immunostaining, AT_1 receptors are specifically localized in hepatic stellate cells, hepatocytes and Kupffer cells with immunoreactivity being much intense in the livers from obese diabetic mice when compared with their respective control livers (Fig. 5.3). With these preliminary data available, we will further address our hypothesis that vitamin D specifically reduces hepatic insulin resistance through modulation of the local hepatic RAS. To address this issue, we will study the effects of hormonal vitamin D on hepatic RAS activity, on hepatic insulin receptors and on post-insulin receptor signalling pathways relevant to insulin sensitivity in vitro and in vivo. We will also determine whether human hepatic insulin sensitivity, specifically, increases with adequate supplementation in hypovitaminosis D and whether this is associated with circulating RAS activity reduction. Insulin resistance contributes to cardiovascular diseases, metabolic syndrome and T2DM, increasingly common world-wide problems presenting huge demand on healthcare resources. Thus, these findings may provide novel approaches to the prevention and management of NAFLD as well as of metabolic syndrome and its sequelae.

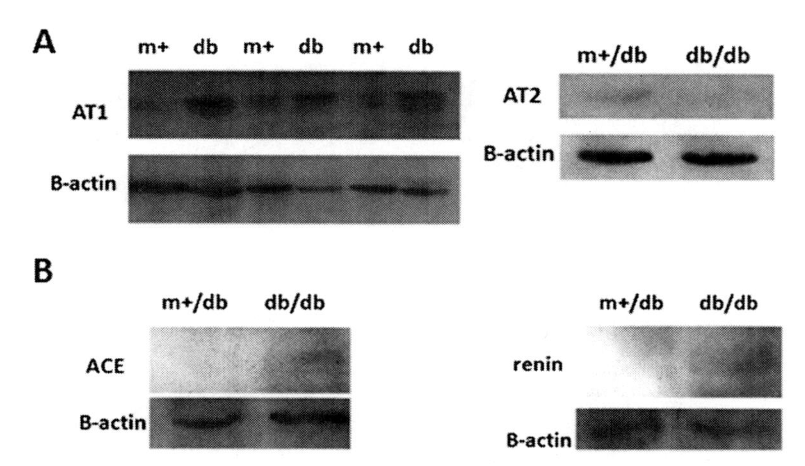

Fig. 5.2 Western blot analysis of some major RAS component expression in the liver of lean m+/db (m+) and diabetic db/db (db) mice. **a** Expression of AT1R protein. Expression of AT2R protein. **b** Expression of ACE protein. Expression of renin protein. The expression levels of AT1R, ACE and renin were upregulated in db/db mouse liver while AT2R had a reduced protein expression when compared with their respective controls

5.4 Local RAS in Intestine

The two primary physiologically roles of intestine are the digestion of food and absorption of electrolytes and nutrients such as glucose. Glucose is a crucial metabolic substrate for most cell types. To achieve glucose uptake by different cell types in our body, there are two categories of glucose transporters, one is the secondary active transporters, i.e. SGLTs and the other is the facilitated transporters, i.e. GLUTs (Table 5.3). The absorption of glucose involves an entry-and-exit mechanism mediated by two separate protein carriers located in the brush border membrane (BBM) and basolateral membrane (BLM) of enterocytes (Wright et al., 1994). This mechanism is energized by a BBM bound ATP-dependent Na^+ pump (SGLT1), an active glucose transporter, that maintains a Na^+ gradient favouring the entry of Na^+ with the concomitant co-transport of glucose or galactose into the enterocytes. SGLT proteins are expressed in many tissues, particularly SGLT1 in the epithelial cells of the small intestine (Wright et al., 2004, 2006) while SGLT2 and SGLT1 are found in the kidney (Pajor et al., 1992; Kanai et al., 1994). A second carrier (GLUT2) is located in the BLM then facilitates glucose expulsion from the cell. The GLUT family of transporters consists of 14 members, but the most commonly expressed isoforms involved in glucose uptake by renal and intestinal epithelia are GLUT1 and GLUT2 (Kellett et al., 2008). At the enterocyte BBM, SGLT1 is a high-affinity, low-capacity transporter that allows active glucose uptake from the intestinal lumen. SGLT1-mediated glucose uptake, in turn, promotes the insertion of GLUT2 into the apical membrane, which provides an additional high capacity pathway for glucose transport that is crucial for absorbing the high level of

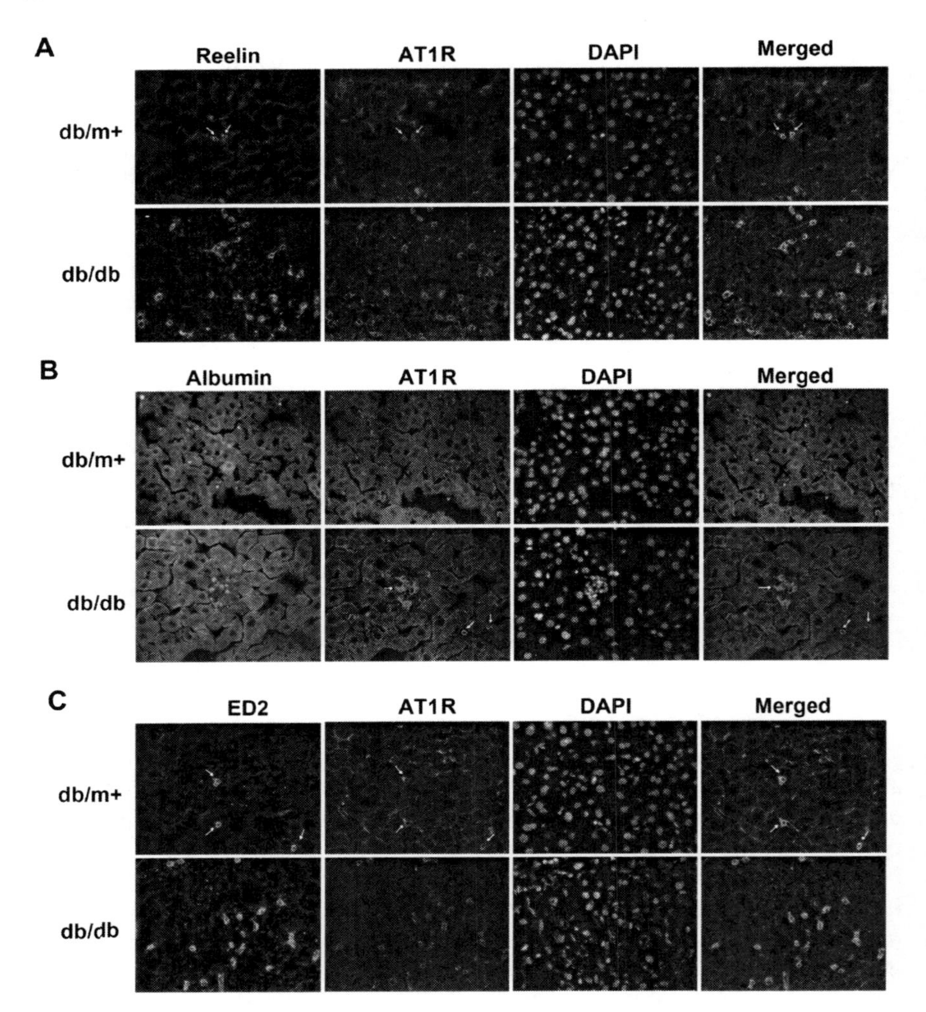

Fig. 5.3 The expression and localization of AT1R in stellate cells (Reelin as a marker) and hepatocytes (albumin marker) and Kupffer cells (ED2 as a marker) from the liver of obese diabetic db/db mice and control lean db/m+ mice. Immunoreactivity of AT1R was immunolabeled with specific liver cell markers of stellate cells, Reelin (**a**), of hepatocytes, Albumin (**b**) and of Kupffer cells, ED2 (**c**). AT1R immunoreactivity was stained with red; Reelin, albumin and ED2 immunoreactivity was stained with green (see *arrows*). Overlay of the immunofluorescence labeling of db/db livers showing more intense immunostaining for AT1R was observed in both stellate cells and Kupffer cells but not in hepatocytes (*merged images*), when compared with their respective control livers. DAPI was used as a marker for cell nuclei (*blue*). Original magnification, ×630 (For interpretation of the references to colour in this figure legend, please be referred to the online version)

luminal glucose generated at peak carbohydrate digestive activity. Indeed, the transport capability of GLUT2 at the BBM can be up to three times greater than that of SGLT1 (Kellett, 2001; Kellett & Brot-Laroche, 2005). Glucose exits the enterocyte

Table 5.3 Different forms of glucose transporters that are involved in active or sodium-dependent (SGLTs) and passive or facilitated (GLUTs) mechanisms. Their locations in tissues and respective function are indicated

Transporter	Tissue	Function
SGLT1	Mainly gut	Intestinal uptake of glucose
SGLT2	Mainly kidney	Renal uptake of glucose
GLUT1	All tissues; mainly red blood cell and brain	Basal uptake of glucose; transport across the blood-brain barrier
GLUT2	Beta-cell of the pancreas, liver, kidney and gut	Regulation of insulin release; other aspects of glucose homeostasis
GLUT3	Brain, kidney, placenta, and other tissues	Uptake into neurons and other tissues
GLUT4	Muscle and adipose tissues	Insulin-mediated uptake of glucose
GLUT5	Gut and kidney	Intestinal absorption of fructose

via GLUT2 or simple diffusion across the BLM. Apart from glucose and galactose, a third form of absorbable monosaccharides is fructose which is transported by another facilitated glucose transporter, GLUT5. In summary, Fig. 5.4a illustrates SGLT1-mediated glucose (as well as galactose) uptake by enterocyte under normal condition. Notwithstanding the involvement of this well-established secondary active transport for glucose uptake, the precise mechanisms by which SGLT1 and GLUT2 are mediated by some novel regulators, and consequently affect glucose uptake at the BBM of enterocytes, have yet to be fully elucidated.

5.4.1 Expression and Function of an Enterocyte RAS

There are diverse factors as well as ever-growing candidates that govern the glucose transporters located at the enterocytes for intestinal glucose uptake. In this regard, the rate of intestinal glucose transport is subject to regulation by both hormones and luminal peptides, which have stimulatory or inhibitory effects. Such stimulatory regulators include insulin (Pennington et al., 1994), pancreatic glucagon (Debnam & Sharp, 1993), gastric inhibitory peptide (Cheeseman & Tsang, 1996), glucagon-like peptide-2 (Cheeseman, 1997), and cholecystokinin (Hirsh & Cheeseman, 1998) as well as leptin (Ducroc et al., 2005) and angiotensin II (Wong et al., 2007; Leung, 2008), the latter two being inhibitory peptides. As far as angiotensin II is concerned, there are several local RASs with functional correlates that have been identified in the gastrointestinal system including, the salivary glands, stomach, intestine, colon, liver, and pancreas (Paul et al., 2006, Leung, 2007). Local RAS components, such as ACE and renin, have been detected in the small intestinal mucosa (Erickson et al., 1992; Seo et al., 1991). It has been proposed that mucosal ACE may function as a brush border membrane peptidase (Yoshioka et al., 1987). Despite the finding that ACE is present at the intestinal BBM, there is no information available on the transport actions of locally produced angiotensin II at this membrane (Stevens et al.,

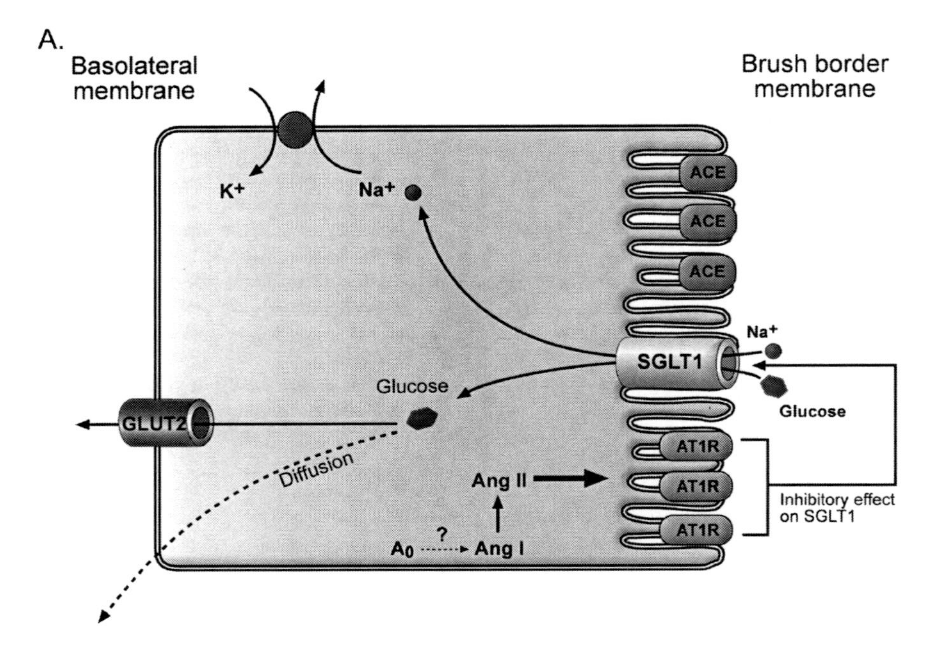

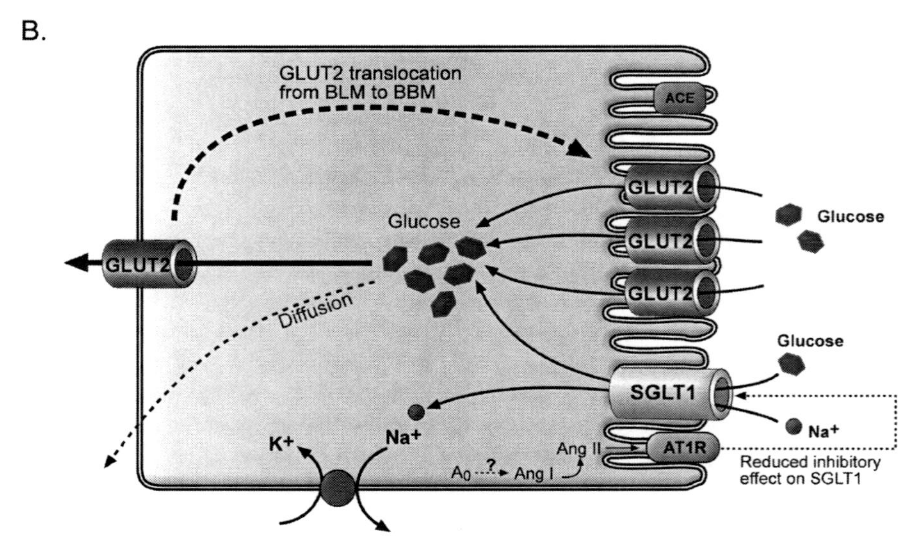

Fig. 5.4 A schematic diagram proposing the potential mechanism of an enterocyte RAS-mediated SGLT1 dependent glucose uptake at the BBM under normal (**a**) and diabetic condition (**b**)

1988; Naim, 1992). Vascular infusion of low concentrations of angiotensin II have been shown to stimulate intestinal fluid transport via the AT_2 receptor; however, higher levels of angiotensin II inhibit fluid transport via an AT_1 receptor-dependent process (Jin et al., 1998). Studies using proximal tubule cells, where the glucose uptake process is very similar to that in enterocytes, imply that locally secreted angiotensin II affects sugar transport. For example, angiotensin II inhibits uptake of the glucose analogue, α-methyl-glucopyranoside into LLC-PK_1 cells and reduces expression of the sodium-dependent glucose transporter at the apical membrane (Kawano et al., 2002). Despite of these findings, direct involvement of a local RAS in the control of intestinal glucose transport, however, has yet to be conclusively demonstrated in the literature.

Against this background, we recently found compelling evidence for the existence of an enterocyte RAS that generates angiotensin II locally and leads to rapid inhibition of SGLT1-mediated intestinal glucose uptake (Wong et al., 2007). In this study, expression of key RAS components at the gene and protein levels were examined in jejunal and ileal enterocytes. Mucosal uptake of glucose by everted intestinal sleeves, before and after addition of angiotensin II to the mucosal buffer, was measured in the presence or absence of losartan, an AT_1 receptor antagonist. The results revealed that enterocytes express angiotensinogen, ACE, and AT_1 and AT_2 receptors; AT_1 receptors and angiotensinogen proteins were specifically localized to the BBM. Expression of angiotensinogen (a mandatory component of a local RAS) and AT_1 and AT_2 receptors, but not ACE, was greater in the ileum than the jejunum. Addition of angiotensin II to the mucosal buffer inhibited phlorizin-sensitive (SGLT1-dependent) jejunal glucose uptake in a rapid and dose-dependent manner, and reduced the expression of SGLT1 at the BBM. Losartan attenuated the inhibitory action of angiotensin II on glucose uptake. Despite this inhibitory effect on glucose uptake, angiotensin II did not affect jejunal uptake of L-leucine, suggesting a specific action on intestinal glucose absorption. The proposed mechanism by which the rapid inhibitory effects of enterocyte-derived angiotensin II is involved in the regulation of SGLT1-mediated intestinal glucose transport at the BBM under normal condition are depicted in Fig. 5.4a.

5.4.2 Enterocyte RAS and Diabetes Mellitus

In light of our enterocyte RAS findings above, we proceed to explore the clinical relevance of this enterocyte RAS in AT_1 receptor-mediated SGLT1-dependent glucose uptake in the intestine. Previous studies have shown that streptozotocin-induced diabetes mellitus promotes glucose transport across the rat intestinal BBM (Hopfer, 1975; Debnam et al., 1988; Burant et al., 1994) and that this is a likely consequence of increased BBM expression of SGLT1 and GLUT2 (Debnam et al., 1988, 1995; Kellett & Brot-Laroche, 2005). Nevertheless, the mechanisms of increased transport remain largely unexplored. Previous studies have shown that proximal tubule cells are able to synthesize and secrete angiotensin II into the luminal fluid, and that angiotensin II might be involved in the regulation of SGLT-mediated transport in these cells (Park & Han, 2001). Interestingly, exposure of proximal tubule cells to high glucose at a level similar to that seen in the plasma of streptozotocin-diabetic animals reduced angiotensin II binding of these cells (Park et al., 2002).

Against this background, we hypothesized that the expression of RAS components at the enterocyte BBM would be subject to regulation by diabetes mellitus and thus affect intestinal glucose uptake (Wong et al., 2009). Our preliminary results showed that streptozotocin-induced diabetes 2 weeks in duration was sufficient to cause a 5-fold increase in blood glucose level, reduce mRNA and protein expression of AT_1- and AT_2-receptors, and reduce ACE levels in isolated jejunal enterocytes. Angiotensinogen expression was, however, stimulated by diabetes whilst renin was not detected in either control or diabetic enterocytes. Considering these data, it is thus plausible to speculate that a local RAS, in the absence of renin, i.e. a renin-independent angiotensin-generating system, exists and is involved in angiotensin II production in the intestine.

On the other hand, streptozotocin-induced diabetes stimulated glucose uptake by 58% and increased the expression of SGLT1 and GLUT2 proteins in purified BBM by 25 and 135%, respectively. Our immunocytochemistry results confirmed that there was an increased BBM expression of GLUT2 in the diabetes condition, which was probably due to the rapid translocation or trafficking of GLUT2 from the BLM to the BBM. Addition of angiotensin II (5 μM), or its non-peptide analogue L-162313 (1 μM), to mucosal buffer decreased glucose uptake by diabetic jejunum by 18 and 24%, respectively. Accordingly, the potency of L-162313 is about 63 times greater than angiotensin II. Furthermore, this inhibitory action of angiotensin II was due to reduced phlorizin-sensitive (SGLT1-dependent) rather than phloretin-sensitive (GLUT2-dependent) transport, and was abolished by the AT_1 receptor antagonist losartan (1 μM). The decreased efficacy of angiotensin II on glucose uptake in diabetes compared to that noted previously in jejunum from normal animals (Wong et al., 2007) is likely to be due to altered RAS expression in diabetic enterocytes, together with disproportionately increased GLTU2 expression at the BBM. Since the rate of intestinal glucose transport affects insulin release and thus glycaemic control in humans, our work raises the issue of whether the enterocyte RAS may be a therapeutic target for reducing glucose transport. On the other hand, the higher potency of L-162313, compared to angiotensin II, in terms of affecting glucose transport, makes it possible that non-peptide AT_1 receptor agonists may be used as gut-directed treatments to reduce postprandial hyperglycaemia in diabetic patients (Wong et al., 2009). In conclusion, The proposed mechanisms by which the release of inhibitory effects of enterocyte-derived angiotensin II may be involved in the regulation of SGLT1-mediated and enhanced trafficking of GLUT2-mediated intestinal glucose transport at the BBM in diabetic intestine (Fig. 5.4b).

5.4.3 ACE2-Angiotensin (1–7)-Mas Receptor Axis and Intestinal Glucose Uptake

Angiotensin converting enzyme 2 (ACE2), the first homologue of ACE2, is a newly identified member of RAS (Bindom & Lazartigues, 2009). Apart from being a cellular receptor for the severe acute respiratory syndrome-associated coronavirus, SARS-CoV (Ren et al., 2006), ACE2 is a membrane-bound mono-carboxypeptidase capable of generating angiotensin (1–9) and angiotensin (1–7) by cleaving the terminal leucine and phenylalanine residue from angiotensin I and angiotensin II, respectively (Iwai & Horiuchi, 2009). As a result there are two separate pathways responsible for angiotensin (1–7) production (Bindom & Lazartigues, 2009). Angiotensin (1–7) is a bioactive peptide of the RAS which has been shown to have beneficial effects on renal and cardiovascular diseases, possibly via the mediation of counter-regulating ACE-angiotensin II-AT1R (Santos et al., 2008). The G-protein-coupled receptor, mas receptor, acts as receptor for angiotensin (1–7). It is found to be expressed in many tissues such as brain, heart, kidney and liver (Iwai & Horiuchi, 2009). In view of this fact, we propose the existence of an ACE2-angiotensin (1–7)-mas axis and its potential role in intestinal glucose uptake. We thus test this hypothesis using a cell model of Caco-2, (Sambuy et al., 2005) a human colon carcinoma cell line by examining the expression and regulation of ACE2-angiotensin (1–7)-mas receptor axis under high glucose exposure.

Our preliminary data shows that components of ACE2, angiotensin (1–7) and mas receptor are expressed in the Caco-2 cells (Fig. 5.5); ACE2 were markedly

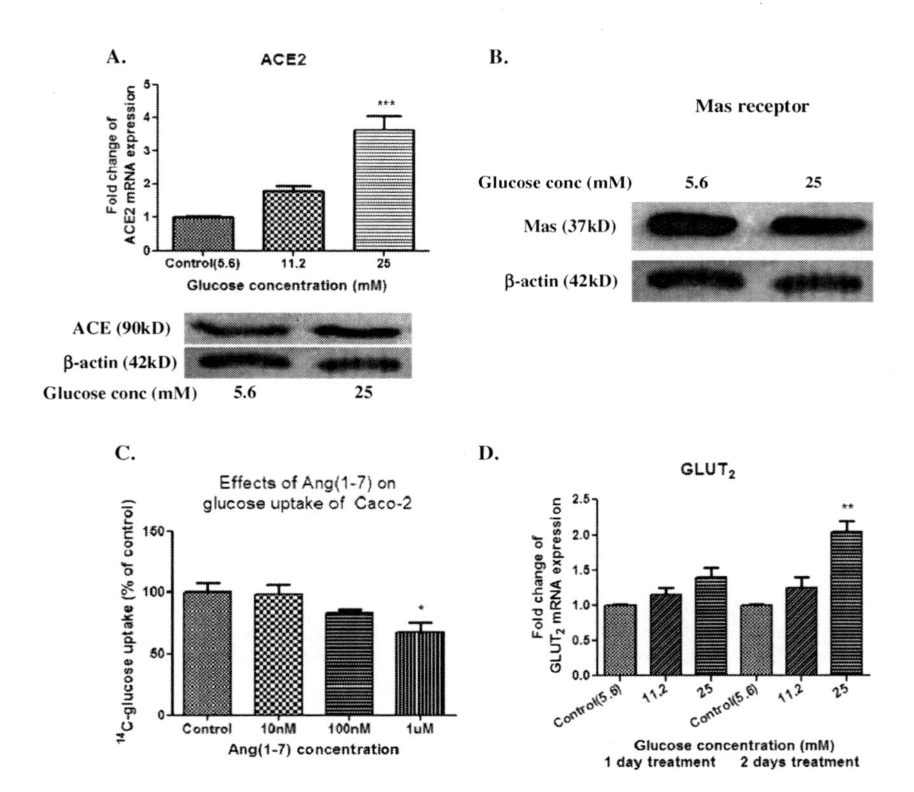

Fig. 5.5 Expression and regulation of ACE2-angiotensin (1–7)-mas receptor under high (25 mM) and normal (5.6 mM) in Caco-2 cell line. **a** ACE2 gene and protein expression. **b** Mas receptor protein expression. **c** ^{14}C-glucose uptake in response to angiotensin (1–7). **d** GLUT2 gene expression

upregulated while being no significant increase in mas receptor under exposure to high glucose when compared with normal concentrations, i.e. 5.6 mM vs. 25 mM (Fig. 5.5a,b). Interestingly, exogenous addition of angiotensin (1–7) appears to dose dependently inhibit glucose uptake by the cell line, as demonstrated by ^{14}C-glucose assay (Fig. 5.5c). On the other hand, GLUT2 gene expression is also dose-dependently upregulated in polarized Caco-2 cells after high glucose exposure. It is significantly upregulated after being exposed to 25 mM glucose for 2 days (Fig. 5.5d). Taken these findings together, our preliminary data indicate the existence of a functional ACE2-angiotensin(1–7)-Mas axis in Caco-2 cell line, which may provide an alternative mechanism for GLUT2 and/or SGLT1-mediated intestinal glucose uptake in normal and diabetic conditions.

References

Albiston AL, McDowall SG, Matsacos D, Simpson RJ, Connolly LM and Chai SY. Evidence that the angiotensin IV (AT(4)) receptor is the enzyme insulin-regulated aminopeptidase. *J Biol Chem* **276**:48623–48626, 2001.

Alison MR. Regulation of hepatic growth. *Physiol Rev* **66**:499–451, 1986.

Allen AM. Angitoensin AT1 receptor-mediated excitation of rat carotid body chemoreceptor afferent activity. *J Physiol* **510**:773–781, 1998.

Arroyo V, Bosch J, Mauri M, Ribera F, Navarro-Lopez F and Rodes J. Effect of angiotensin-II blockade on systemic and hepatic haemodynamics and on the renin-angiotensin-aldosterone system in cirrhosis with ascites. *Eur J Clin Invest* **11**:221–229, 1981.

Atzori L, Poli G and Perra A. Hepatic stellate cell: a star cell in the liver. *Int J Biochem Cell Biol* **41**:1639–1642, 2009.

Bataller R, Gines P, Nicolas JM, Gorbig MN, Garcia-Ramallo E, Gasull X, Bosch J, Arroyo V and Rodes J. Angiotensin II induces contraction and proliferation of human hepatic stellate cells. *Gastroenterology* **118**:1149–1156, 2000.

Bindom SM and Lazartigues E. The sweeter side of ACE2: physiological evidence for a role in diabetes. *Mol Cell Endocrinol* **302**:193–202, 2009.

Boucher BJ. Inadequate vitamin D status: does it contribute to the disorders comprising syndrome 'X'? *Br J Nutr* **79**:315–327, 1998.

Burant CF, Flink S, DePaoli AM, Chen J, Lee WS and Chang EB. Small intestinal hexose transport in experimental diabetes: increased transporter mRNA and protein expression in enterocytes. *J Clin Invest* **93**:578–585, 1994.

Burgess ED, Hawkins RG and Watanabe M. Interaction of 1,25-dihydroxyvitamin D and plasma renin activity in high renin essential hypertension. *Am J Hypertens* **3**:903–905, 1990.

Casini A. Alcohol-induced fatty liver and inflammation: where do Kupffer cells act? *J Hepatol* **32**:1026–1030, 2000.

Chan H, Leung PS and Tam SC. Effect of angiotensin AT1 receptor antagonist on D-galactosamine-induced acute liver injury. *Clin Exp Pharmacol Physiol* **34**:985–991, 2007.

Cheeseman CI. Upregulation of SGLT-1 transport activity in rat jejunum induced by GLP-2 infusion in vivo. *Am J Physiol* **273**:R1965–R1971, 1997.

Cheeseman CI and Tsang R. The effect of GIP and glucagon-like peptides on intestinal basolateral membrane hexose transport. *Am J Physiol* **271**:G477–G482, 1996.

Debnam ES, Smith MW, Sharp PA, Srai SK, Turvey A and Keable SJ. The effects of streptozotocin diabetes on sodium-glucose transporter (SGLT1) expression and function in rat jejunal and ileal villus-attached enterocytes. *Eur J Physiol* **430**:151–159, 1995.

Debnam ES and Sharp PA. Acute and chronic effects of pancreatic glucagon on sugar transport across the brush-border and basolateral membranes of rat jejunal enterocytes. *Exp Physiol* **78**:197–207, 1993.

Debnam ES, Karasov WH and Thompson CS. Nutrient uptake by rat enterocytes during diabetes mellitus: evidence for an increased sodium electrochemical gradient. *J Physiol* **397**: 503–512, 1988.

Ducroc R, Guilmeau S, Akasbi K, Devaud H, Buyse M and Bado A. Luminal leptin induces rapid inhibition of active intestinal absorption of glucose mediated by sodium-glucose cotransporter 1. *Diabetes* **54**:348–354, 2005.

Erickson RH, Suzuki Y, Sedlmayer A, Song IS and Kim YS. Rat intestinal angiotensin-converting enzyme: purification, properties, expression, and function. *Am J Physiol* **263**:466–473, 1992.

Fung ML, Lam SY, Wong TP, Tjong YW and Leung PS. Carotid body AT4 receptor expression and its upregulation in chronic hypoxia. *Open Cardiovasc Med J* **1**:1–7, 2007.

Fung ML, Lam SL, Dong X, Chen Y and Leung PS. Postnatal hypoxemia increases angiotensin II sensitivity and upregulates AT_{1a} angiotensin receptors in rat carotid body chemoreceptors. *J Endocrinol* **173**:305–313, 2002.

Fung ML, Lam SY, Dong X, Chen Y and Leung PS. Functional expression of angiotensin II receptors in type-I cells of the rat carotid body. *Eur J Endocrinol* **441**:474–480, 2001.

Gascon-Barre M, Demers C, Mirshahi A, Neron S, Zaizal S and Nanci A. The normal liver harbors the vitamin D nuclear receptor in non-parenchymal and biliary epithelial cells. *Hepatology* **37**:1034–1042, 2003.

Gonzalez C, Almaraz L, Obeso A and Rigual R. Carotid body chemoreceptors: from natural stimuli to sensory discharges. *Physiol Rev* **74**:829–898, 1994.

Hardikar W and Suchy FJ. Hepatobiliary function. In Boron WF and Boulpaep EL (eds), Medical physiology, pp 975–1002, Saunders, New York, NY, 2003.

Harrison SA. New treatments for nonalcoholic fatty liver disease. *Curr Gastroenterol Rep* **8**: 21–29, 2006.

Herath CB, Warner FJ, Lubel JS, Dean RG, Jia Z, Lew RA, Smith AI, Burrell LM and Angus PW. Upregulation of hepatic angiotensin-converting enzyme 2 (ACE2) and angiotensin-(1–7) levels in experimental biliary fibrosis. *J Hepatol* **47**:387–395, 2007.

Hirose A, Ono M, Saibara T, Nozaki Y, Masuda K, Yoshioka A, Takahashi M, Akisawa N, Iwasaki S, Oben JA and Onishi S. Angiotensin II type 1 receptor blocker inhibits fibrosis in rat nonalcoholic steatohepatitis. *Hepatology* **45**(6):1375–1381, 2007.

Hollis BW. 25-Hydroxyvitamin D3-1 alpha-hydroxylase in porcine hepatic tissue: subcellular localization to both mitochondria and microsomes. *Proc Natl Acad Sci USA* **87**: 6009–6013, 1990.

Hopfer U. Diabetes mellitus: changes in the transport properties of isolated intestinal microvillous membranes. *Proc Nat Acad Sci USA* **72**:2027–2031, 1975.

Iwai M and Horiuchi M. Devil and angel in the renin-angiotensin system: ACE-angiotensin II-AT1 receptor axis vs. ACE2-angiotensin-(1–7)-Mas receptor axis. *Hypertens Res* **32**: 533–536, 2009.

Jin XH, Wang ZQ, Siragy HM, Guerrant RL and Carey RM. Regulation of jejunal sodium and water absorption by angiotensin subtype receptors. *Am J Physiol* **275**:R515–R523, 1998.

Kanai Y, Lee W-S, You G, Brown D and Hediger MA. The human kidney low affinity Na^+/glucose cotransporter SGLT2. *J Clin Invest* **93**:397–404, 1994.

Kawano K, Ikari A, Nakano M and Suketa Y. Phosphatidylinositol 3-kinase mediates inhibitory effect of angiotensin II on sodium/glucose cotransporter in renal epithelial cells. *Life Sci* **71**: 1–13, 2002.

Kellett GL, Brot-Laroche E, Mace OJ and Leturque A. Sugar absorption in the intestine: the role of GLUT2. *Ann Rev Nutr* **28**:1–20, 2008.

Kellett GL and Brot-Laroche E. Apical GLUT2: a major pathway of intestinal sugar absorption. *Diabetes* **54**:3056–3062, 2005.

Kellett GL. The facilitated component of intestinal glucose absorption. *J Physiol* **531**:585–595, 2001.

Kong J and Li YC. Effect of ANG II type I receptor antagonist and ACE inhibitor on vitamin D receptor-null mice. *Am J Physiol* **285**:R255–R261, 2003.

Lam SY, Fung ML and Leung PS. Regulation of the angiotensin-converting enzyme activity by a time-course hypoxia in the carotid body. *J Appl Physiol* **96**:809–813, 2004.

Lam SY and Leung PS. Chronic hypoxia activates a local angiotensin-generating system in rat carotid body. *Mol Cell Endocrinol* **203**:147–153, 2003.

Lam SY and Leung PS. A locally generated angiotensin system in rat carotid body. *Regul Pept* **107**:97–103, 2002.

Lavoie JL and Sigmund CD. Overview of the renin-angiotensin system: an endocrine and paracrine system. *Endocrinology* **144**:2179–2183, 2003.

Leung PS. Angiotensin II and intestinal glucose uptake. *Physiol News* **72**:PN21, 2008 (www.physoc.org).

Leung PS. The physiology of a local renin-angiotensin system in the pancreas. *J Physiol* **580**: 31–37, 2007.

Leung PS. Novel roles of a local angiotensin-generating system in the carotid body. *J Physiol* **575**:4, 2006.

Leung PS. The peptide hormone angiotensin II: its new functions in tissues and organs. *Curr Protein Pept Sci* **5**:267–273, 2004.

Leung PS, Fung ML and Tam MS. Renin-angiotensin system in the carotid body. *Int J Biochem Cell Biol* **35**(6):847–854, 2003.

Leung PS, Lam SY and Fung ML. Chronic hypoxia upregulates the expression and function of AT_1 receptor in rat carotid body. *J Endocrinol* **167**:517–524, 2000.

Li YC, Kong J, Wei M, Chen ZF, Liu SQ and Cao LP. 1,25-Dihydroxyvitamin D(3) is a negative endocrine regulator of the renin-angiotensin system. *J Clin Invest* **110**:229–238, 2002.

Li Y and Schultz HD. Enhanced sensitivity of Kv channels to hypoxia in the rabbit carotid body in heart failure: role of angiotensin II. *J Physiol* **575**:215–227, 2006.

Lind L, Hanni A, Lithell H, Hvarfner A, Sorensen OH and Ljunghall S. Vitamin D is related to blood pressure and other cardiovascular risk factors in middle-aged men. *Am J Hypertens* **8**:894–901, 1995.

Lubel JS, Herath CB, Tchongue J, Grace J, Jia Z, Spencer K, Casley D, Crowley P, Sievert W, Burrell LM and Angus PW. Angioensin-(1–7), an alternative metabolite of the renin-angiotensin system, is up-regulated in human liver disease and has anti-fibrotic activity in the bile-duct-ligated rat. *Clin Sci* **117**:375–386, 2009.

Lubel JS, Herath CB, Burrell LM and Angus PW. Liver disease and the renin-angiotensin system: recent discoveries and clinical implications. *J Gastroenterol Hepatol* **23**:1327–1338, 2008.

Naim HY. Angiotensin-converting enzyme of the human small intestine. Subunit and quaternary structure, biosynthesis and membrane association. *Biochem J* **286**:451–457, 1992.

Paizis G, Tikellis C, Cooper ME, Schembri JM, Lew RA, Smith AI, Shaw T, Warner FJ, Zuilli A, Burrell LM and Angus PW. Chronic liver injury in rats and humans upregulates the novel enzyme angiotensin converting enzyme 2. *Gut* **54**:1790–1796, 2005.

Pajor AM, Hirayama BA and Wright EM. Molecular evidence for two renal Na^+/glucose cotransporters. *Biochem Biophys Acta* **1106**:216–220, 1992.

Park SH and Han HJ. The mechanism of angiotensin II binding downregulation by high glucose in primary renal proximal tubule cells. *Am J Physiol* **282**:F228–F237, 2001.

Park SH, Woo CH, Kim JH, Lee JH, Yang IS, Park KM and Han HJ. High glucose down-regulates angiotensin II binding via PKC-MAPK-$cPLA_2$ signal cascade in renal proximal tubule cells. *Kidney Int* **61**:913–925, 2002.

Paul M, Mehr AP and Kreutz R. Physiology of local renin-angiotensin systems. *Physiol Rev* **86**:747–803, 2006.

Pennington AM, Corpe CP and Kellett GL. Rapid regulation of rat jejunal glucose transport by insulin in a luminally and vascularly perfused preparation. *J Physiol* **478**:187–193, 1994.

Poli G. Pathogenesis of liver fibrosis: role of oxidative stress. *Mol Aspects Med* **21**:49–98, 2000.

Ren X, Glende J, Al-Falah M, de Vries V, Schwegmann-Wessels C, Qu X. Analysis of ACE2 in polarized epithelial cells: surface expression and function as receptor for severe acute respiratory syndrome-associated coronavirus. *J Gen Virol* **87**:1691–1695, 2006.

Rockey DC and Weisiger RA. Endothelin induced contractility of stellate cells from normal and cirrhotic rat liver: implications for regulation of portal pressure and resistance. *Hepatology* **24**:233–240, 1996.

Sambuy Y, De Angelis I, Ranaldi G, Scarino ML, Stammati A and Zucco F. The Caco-2 cell line as a model of the intestinal barrier: influence of cell and culture-related factors on Caco-2 cell functional characteristics. *Cell Biol Toxicol*. **21**:1–26, 2005.

Santos RA, Ferreira AJ, Simoes E and Silva AC. Recent advances in the angiotensin-converting enzyme 2-angiotensin(1–7)-Mas axis. *Exp Physiol* **93**:519–527, 2008.

Schneider AW, Kalk JF and Klein CP. Effect of losartan, an angiotensin II receptor antagonist, on portal pressure in cirrhosis. *Hepatology* **29**:334–339, 1999.

Seo MS, Fukamizu A, Saito T and Murakami K. Identification of a previously unrecognized production site of human renin. *Biochim Biophys Acta* **1129**:87–89, 1991.

Stevens BR, Fernandez A, Kneer C, Cerda JJ, Phillips MI and Woodward ER. Human intestinal brush border angiotensin-converting enzyme activity and its inhibition by antihypertensive Ramipril. *Gastroenterology* **94**:942–947, 1988.

Targher G, Bertolini L, Scala L, Ciglioni M, Zenari L, Falezza G and Arcaro G. Associations between serum 25-hydroxyvitamiun D3 concentrations and liver histology in patients with non-alcoholic fatty liver disease. *Nutr Metab Cardiovasc Dis* **17**:517–524, 2007.

Wang TT, Tavera-Mendoza LE, Lapierre D, Libby E, Burton MacLeod NB, Nagai Y, Bourdeau V, Konstorum A, Lallemant B and Zhang R. Large-scale in silico and microarray-based identification of direct 1,25-dihydroxyvitamin D_3 target genes. *Mol Endocrinol* **19**:2685–2895, 2005.

Wong TP, Debnam ES and Leung PS. Diabetes mellitus and expression of the enterocyte renin-angiotensin system: implications for control of glucose transport across the brush border membrane. *Am J Physiol* **297**:C601–C610, 2009.

Wong TP, Debnam ES and Leung PS. Involvement of an enterocyte renin-angiotensin system in the local control of SGLT1-dependent glucose uptake across the rat small intestinal brush border membrane. *J Physiol* **584**:613–623, 2007.

Woo J, Lam CW, Leung J, Lau WY, Lau E, Ling X, Xing X, Zhao XH, Skeaff CM, Bacon CJ, Rockell JE, Lambert A, Whiting SJ and Green TJ. Very high rates of vitamin D insufficiency in women of child-bearing age living in Beijing and Hong Kong. *Br J Nutr* **99**:1330–1334, 2008.

Wright EM, Hirayama BA and Loo DF. Active sugar transport in health and disease. *J Int Med* **261**:32–43, 2006.

Wright EM, Loo DDF, Hirayama BA and Turk E. Surprising versatility of Na^+/glucose cotransporters (SLC5). *Physiology* **19**:370–376, 2004.

Wright EM, Hirayama BA, Loo DDF, Turk E and Hager K. Intestinal sugar transport. In LR Johnson (ed), Physiology of the gastrointestinal tract, 3rd edn. Raven Press, New York, NY, pp 1751–1772, 1994.

Yoshiji H, Kuriyama S, Yoshii J, Ikenaka Y, Noguchi R, Nakatani T, Tsujinoue H and Fukui H. Angiotensin-II type 1 receptor interaction is a major regulator for liver fibrosis development in rats. *Hepatology* **34**:745–750, 2001.

Yoshioka M, Erickson RH, Woodley JF, Gulli R, Guan D and Kim YS. Role of rat intestinal brush-border membrane angiotensin-converting enzyme in dietary protein digestion. *Am J Physiol* **253**:G781–G786, 1987.

Chapter 6
Pancreatic RAS

Metabolic syndrome disorder is a grave health problem affecting diverse populations worldwide. Among them, type 2 diabetes mellitus (T2DM) is a particularly widespread cardiometabolic syndrome which is characterized by impaired peripheral insulin sensitivity and pancreatic β-cell function. Its prevalence is ever increasing, not only in adults, but also notably in young people. Despite ongoing efforts to generally improve people's dietary habits and exercise-related behaviour, T2DM has become an epidemic beyond preventability in many individuals. It is also recognized that, if allowed to develop, the cardiovascular and renal sequelae of T2DM will become a major financial burden on nations' healthcare systems. Hence, we are faced with an urgent need for measures to prevent or stall disease progression. In this regard, mechanistic insight into the regulatory pathways that may be harnessed to preserve the structure and function of pancreatic islets in affected persons represents a promising approach. Knowledge garnered from elucidating these basic mechanisms can be translated into clinical applications and offers prospects for rapid development of new therapeutic strategies to limit the health impacts of T2DM and its cardiovascular complications (Garber, 2009).

A recent meta-analysis of 12 randomized clinical trials showed that pharmacological inhibition of the renin-angiotensin system (RAS) protects against T2DM onset (Abuissa et al., 2005). The protective mechanisms are believed to include reduced insulin resistance of peripheral tissues; however, the beneficial effects of RAS blockade in enhancing pancreatic islet function and structure remain ambiguous. Apart from the circulating RAS, multiple cellular and tissue RASs have been shown, in the last decade, to exhibit RAS activities in the regulation of tissue and organ functions (Paul et al., 2006). Of great interest in this context is the recently identified local RAS in the pancreas which has novel roles complementing, counteracting, and/or independent of the circulating RAS. The proposed roles include an array of physiological actions, such as cell proliferation, apoptosis, reactive oxygen species (ROS) generation, inflammation, fibrosis, and hormonal secretion. Dysregulation of the pancreatic RAS may result in a variety of pancreatic and metabolic diseases arising from exocrine and endocrine glands (Leung, 2007). In this chapter, evidence for the existence of a local RAS in the pancreas, as

P.S. Leung, *The Renin-Angiotensin System: Current Research Progress in The Pancreas*, Advances in Experimental Medicine and Biology 690,
DOI 10.1007/978-90-481-9060-7_6,

demonstrated by the localization and expression studies of major RAS components from different pancreatic cell types, namely acinar, ductal, stellate and islet cells, will be critically reviewed.

6.1 Acinar Cell RAS

The acinar cell is the predominant cell type of exocrine pancreas, constituting ~80% of the whole gland. The primary function of this secretory cell is to produce digestive enzymes (see Chapter 2). Malfunction of pancreatic acinar cells is closely associated with pancreatic disease, notably pancreatic inflammation. Previous studies have shown that pancreatic inflammation was concomitantly accompanied by pancreatic apoptosis and fibrosis, ultimately leading to cell damage and injury, as observed in humans and animal models of chronic pancreatitis (CP); indeed, acinar cell apoptosis is a key event in pancreatic fibrosis, which in turn results in pancreatic atrophy (Bateman et al., 2002; Hashimoto et al., 2000). In the last decade, substantial experimental data have emerged supporting the existence of acinar RAS in the pancreas, most notably demonstrations in vitro and in vivo of the expression and localization of major RAS components.

One particular rat cell line (AR42J cells) has commonly been employed in examinations of RAS component expression in acinar cells. The pancreatic acinar AR42J cell is a stable rat pancreatic tumour cell line with demonstrated qualities and functions characteristic of acinar cells in vivo, such as the ability to synthesize and secrete pancreatic enzymes like trypsin, amylase, chymotrypsin, kallikrein and cholesterol esterase. In addition, critical peptide hormone receptors, such as cholecystokinin (CCK), have been identified in AR42J cells (Chappell et al., 1995; Christophe, 1994). In view of these properties, it is a useful and even an indispensable cell model in which to study the physiology and pathophysiology of acinar cells in the field of pancreatology.

Of great interest in this context is the fact that all key RAS elements have been identified and characterized in this cell line, including renin, angiotensin-converting enzyme (ACE), angiotensinogen, angiotensin II, and other active angiotensin peptides, as well as their respective receptors (Chappell et al., 2001). Chappell and his colleagues first reported that the AR42J cell line expresses binding sites for angiotensin II. The membrane of the AR42J cell was found to display various binding affinities for angiotensin peptides, with angiotensin II being the most predominant [angiotensin II $\geq$ angiotensin III $\gg$ angiotensin I > angiotensin (1–7) $\gg$ angiotensin (1–6)] (Chappell et al., 1992, 1995). Using ^{125}I-[Sar^1, Thre^8]-angiotensin II, studies of distribution sites have shown a high density of angiotensin II binding sites in pancreatic acinar cells. Interestingly, the most predominant binding sites for angiotensin II in AR42J cells are AT_2 receptors, with a molecular weight of approximately 110 kDa, which is greater than that based on the analysis of AT_2 receptor protein sequence. Specific antagonism of the AT_2 receptor, but not the AT_1 receptor, abolished angiotensin II binding activities (Chappell et al., 1995, 2001).

Some angiotensin II binding sites on AR42J cells were also sensitive to AT_1 receptor antagonism, however, indicating that there are AT_1 receptors on the cells.

There are two isoforms of AT_1 receptor, namely AT_{1a} and AT_{1b} receptor subtypes that are expressed in AR42J cells but the AT_{1a} receptor subtype was found to be the predominant isoform on AR42J cells, as evidenced by RT-PCR analysis (Leung & Chappell, 2003). It has been shown that AT_1 receptors generally mediate angiotensin II-induced changes in intracellular Ca^{2+} levels by activation of phospholipase C, through a pertussis toxin-insensitive pathway as well as stimulating inositol 1,4,5-trisphosphate (IP_3) and influencing cellular Ca^{2+} stores, thus finally eliciting biological action of angiotensin II (Yin et al., 2003). In consistent with our expression data, AT_{1a} receptor has also been confirmed to be the predominant subtype of AT_1 receptor expressed in the AR42J cells (Cheung et al., 1999). More importantly, a functional AT_{1a} receptor has been characterized in AR42J cells and angiotensin II were found to mediate AT_1 receptor activation and thus produce a dose-dependent increase in α-amylase release and inositol phosphate production from the cell line; indeed, high-performance liquid chromatography analysis showed that angiotensin II stimulates the rapid accumulation of IP_3 (Cheung et al., 1999). In this study, it is intriguing that angiotensin II-induced α-amylase secretion could be blocked by losartan, a specific antagonist for AT_1 receptor while CGP42112, a selective agonist for AT_2 receptor, did not exhibit an appreciable effect on amylase release at micromolar concentrations (Cheung et al., 1999). In addition to angiotensin II receptors, renin, angiotensinogen and ACE along with angiotensin II, angiotensin III and angiotensin (1–7), were expressed in the pancreatic acinar AR42J cell line, as demonstrated by RT-PCR analysis and radio-immunoassay coupled with high-performance liquid chromatography (Leung & Chappell, 2003). Table 6.1 provides a summary of the key RAS components that are expressed in the rat AR42J cell line as well as acinar cells from different animal species.

Apart from the pancreatic rat acinar cell line, expression of acinar RAS components has also been detected in the whole pancreas and isolated acinar cells

Table 6.1 The expression of RAS components in acinar cell line and acinar cells from various animal species. "+" and "–" denote the presence and absence of the respective component, respectively

RAS components/cell types	Renin	ACE	Angiotensinogen	Angiotensin II	AT_1 receptor	AT_2 receptor
Rat AR42J cell	+	+	+	+	+	+
Canine acinar cell	–	–	+	+	+	+
Rat acinar cell	–	+	+	+	+	+
Mouse acinar cell	–	–	–	+	–	–
Human acinar cell	–	–	+	–	+	–

from various animal species. Angiotensinogen mRNA and protein, angiotensin II, angiotensin III, angiotensin (1–7), and high affinity binding sites for angiotensin II were found to be present in dog pancreas (Chappell et al., 1991). Amongst the angiotensin peptides, angiotensin II is the most abundant bioactive peptide, with concentrations present at approximately fivefold those of angiotensin (1–7). Intriguingly, the concentrations of these three angiotensin peptides in the pancreatic tissue levels are three-fold higher than those present in the blood circulation. Specific binding sites for ^{125}I-angiotensin II were observed overlying the pancreatic acinar cells, and arteriolar and duct tissues, whereas angiotensin I and renin-like activity were not detected in the exocrine pancreas (Chappell et al., 1991). AT_1 receptor and AT_2 receptor immunoreactivity was previously localized to pancreatic acinar cells in mice and rats, with staining intensity being much weaker than that observed in the pancreatic duct cells of the exocrine pancreas (Leung et al.,

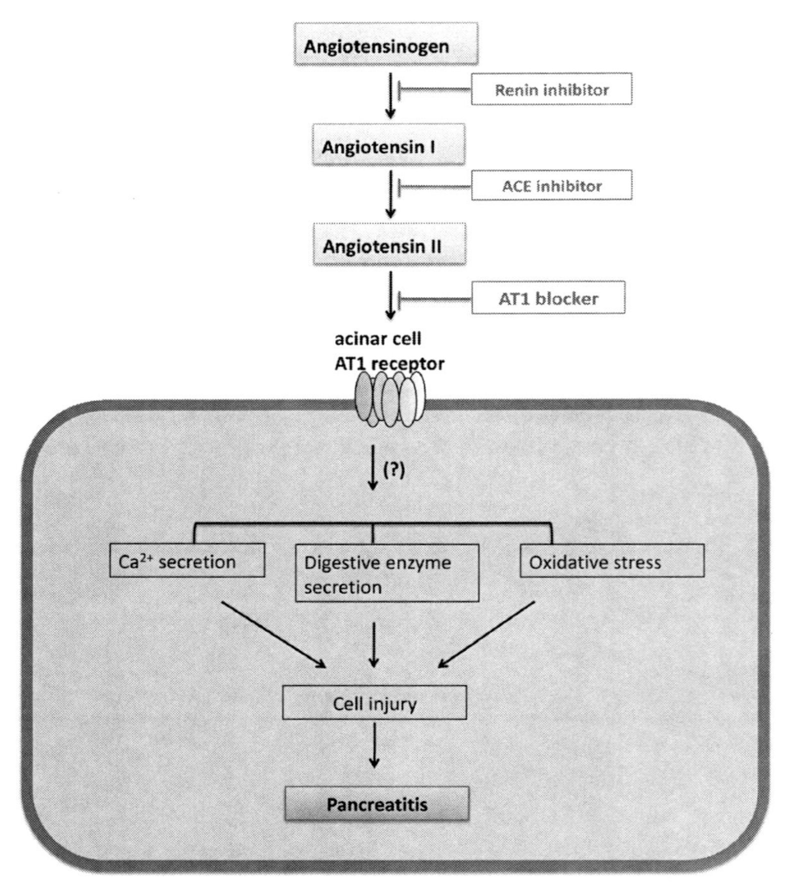

Fig. 6.1 A proposed existence and functional implications of a pancreatic acinar cell RAS in the pancreas

1997a). In addition, angiotensin II immunoreactivity was localized predominantly to ductal epithelium and there was less pronounced immunoreactivity in the acinar cells of mouse exocrine pancreas (Leung et al., 1998). On the other hand, RT-PCR experiments revealed angiotensinogen, AT_1, and AT_2 receptors in isolated rat pancreatic acinar cells (Tsang et al., 2004). In that study, exogenous addition of angiotensin II stimulated a dose-dependent secretion of digestive enzymes from acinar cells (Tsang et al., 2004), which is consistent with findings in pancreatic acinar AR42J cells (Cheung et al., 1999). In the human pancreas, angiotensinogen and AT_1 immunoreactivities were localized in the ductal epithelium of exocrine pancreas; the precise localization of RAS components in acinar cells have yet to be determined (Lam & Leung, 2002).

As far as clinical setting is concerned, acinar RAS components are sensitive to pathophysiological conditions, such as pancreatitis and hypoxia (Leung & Chappell, 2003). Figure 6.1 presents a schematic representation proposing the existence of acinar RAS and its potential functional correlates in the exocrine pancreas. Further discussion on the research progress on the RAS in exocrine pancreas concerning pancreatic inflammation will be critically reviewed in Chapter 8.

6.2 Ductal Cell RAS

The pancreatic duct cell is a minor cell type in the exocrine pancreas, comprising only a small portion of the organ: 14% by volume in human, 4% in guinea pig, and 2% in rat (Githens, 1988). The duct cell has three primary functions: (1) it provides a structural scaffold for acinar cells; (2) it facilitates the flow of digestive enzymes to the duodenum; and (3) it produces sodium bicarbonate for the neutralization of gastric chyme. Defects in ductal function may be attributed to various pancreatic diseases, such as cystic fibrosis, duodenal ulcer, pancreatitis, and pancreatic cancer (more than 90% are ductal in origin).

Previous studies have reported the expression and localization of major RAS components in pancreatic duct cells, which may be relevant for the physiology and pathophysiology of the exocrine pancreatic duct system. Firstly, angiotensinogen, the mandatory component of a local RAS, has been demonstrated to be expressed and localized predominantly in the epithelial cells of pancreatic ducts, as well as in the endothelial lining of pancreatic blood vessels (Leung et al., 1999). Interestingly, angiotensinogen was also detected in significant amounts in pancreatic juice from rats; in contrast, rat duct cells did not express angiotensinogen at an immunoassay detectable level. Instead, most of the angiotensinogen-immunoreactive cells were from glucagon-secreting cells (Regoli et al., 2003). In the human pancreas, immunoreactivity of angiotensinogen was found in the cytoplasm of the epithelial cells from pancreatic ductal epithelium, and was upregulated in pancreatic endocrine tumour tissue (Lam & Leung, 2002). The presence of angiotensinogen in duct cells indicates that locally produced angiotensin II should be present in the pancreatic duct. Indeed, angiotensin II immunoreactivity was observed in the ductal epithelial cells of medium-sized pancreatic ducts throughout mouse pancreatic tissue (Leung et al., 1998). Moreover, ^{125}I-labeled angiotensin II binding was observed in canine pancreatic duct tissue (Chappell et al., 1991), with AT_2 receptors being the predominant receptors (Chappell et al., 1992). In murines, both AT_1 and AT_2 receptors were predominantly localized in the ductal epithelium (Leung et al., 1997a). In human, AT_1 receptor was also expressed in the pancreatic ducts (Lam & Leung, 2002); ACE expression was detected in the connective tissues surrounding the pancreatic ducts and was co-localized with AT_1 receptor in human pancreas (Arafat et al., 2007). Furthermore, several RAS components are consistently observed in pancreatic duct cell lines. For example, CFPAC-1 cells, commonly used in cystic fibrosis studies, were found to express AT_1 receptors, thus implicating its role for angiotensin II-induced ductal ion transport by the duct cells (Chan et al., 1997; Cheng et al., 1999). AT_1 receptors have also been found in the cell membranes and, occasionally, the cytoplasm at various levels of expression in several pancreatic ductal adenocarcinoma cell lines; their expression was sensitive to regulation by angiotensin II via mediation of the AT_1 receptor (Anandanadesan et al., 2008).

As far as functional correlate is concerned, previous studies have suggested a novel role for angiotensin II in the regulation of ion transport from the secretory epithelia of a number of tissues and organs such as the intestine (Cox et al., 1987), trachea (Norris et al., 1991), and epididymis (Leung et al., 1997b). In the intestine, angiotensin II was recently shown to have a direct role in regulating intestinal epithelial Na^+ transport via enhanced activity of an intestinal Na^+–H^+ exchanger, and thereby controlling electrolyte and fluid absorption (Musch et al., 2009). In the pancreas, ductal RAS has also been shown to have a physiological role in the regulation of ductal cell anion secretion from the pancreatic duct, such that the balance between RAS stimulation and inhibition governs pancreatic ductal HCO_3^- secretion. In this regard, angiotensin II was found to induce changes in intracellular Ca^{2+} and cAMP, thus stimulating Cl^- secretion in the cystic fibrosis pancreatic duct cell line CFPAC-1 (Cheng et al., 1999). Selective antagonism of AT_1 receptors in CFPAC-1 cells by losartan inhibited Cl^- secretion across a CFPAC-1 monolayer. Interestingly, this AT_1 receptor-mediated anion secretion was polarity specific, dependent on the application of losartan to the apical or basolateral membranes. These data are consistent with an immunohistochemical study showing differential expression of AT_1 receptors in apical and basolateral membranes of mouse pancreatic duct (Chan et al., 1997). Taken together, these data demonstrate that angiotensin II can stimulate Cl^- secretion in human duct cells by activating apical Cl^- channels via AT_1 receptors in a manner that depends on Ca^{2+} and cAMP (Cheng et al., 1999). On the other hand, it has been reported that AT_1 receptor activation-induced increases in calcium affect calcium-mediated chloride channels in dog pancreatic duct epithelial cells and CFPAC-1 cells, further suggesting that angiotensin II may play a role in regulating pancreatic ductal secretion (Fink et al., 2002). Interestingly, angiotensin II can enhance secretin-induced pancreatic secretion while the ACE inhibitor, captopril, can inhibit this pancreatic bicarbonate output from the pancreas (Howard-McNatt & Fink, 2002).

About 95% of exocrine pancreatic cancer cases involve adenocarcinomas originating from the pancreatic ducts. Pancreatic ductal adenocarcinoma (PDA) accounts for more than 20% of the mortality rate of gastrointestinal malignancies (Korc, 2003). It is known that vascular endothelial growth factor (VEGF) is an important pro-angiogenic mediator for PDA; it is implicated in tumour progression and resistance, enhancing proliferation and survival of tumour cells (Korc, 2003). In this context, ACE and AT_1 receptor were localized to ductal epithelial cells in PDA specimens. Angiotensin II is a potent stimulator of VEGF expression in PDA and this effect is directly mediated by AT_1 receptors, which in turn are dependent on tyrosine kinase activity and ERK1/2 MAP kinase activation; in this case, AT_1 receptor blocker and ACE inhibitor could decrease VEGF mRNA and protein expression and thus inhibit proliferation of PDA tumour cells (Anandanadesan et al., 2008). Moreover, co-localization of the AT_1 receptor and VEGF was observed in ductal cells of normal human pancreas, indicating that there may be paracrine and/or autocrine actions of angiotensin II and VEGF on angiogenesis. Furthermore, angiotensin II is also a potential stimulus for monocyte chemoattractant protein-1 (MCP-1) in PDA cells mediated by AT_1 receptor-ERK1/2-NFkB dependent mechanism (Chehl et al., 2009). All these data suggest that RAS antagonism may help reduce tumour growth, angiogenesis and metastasis (Fujita et al., 2002). In summary, Fig. 6.2 provides a schematic representation proposing the existence of ductal RAS and its potential functional correlates in the pancreas.

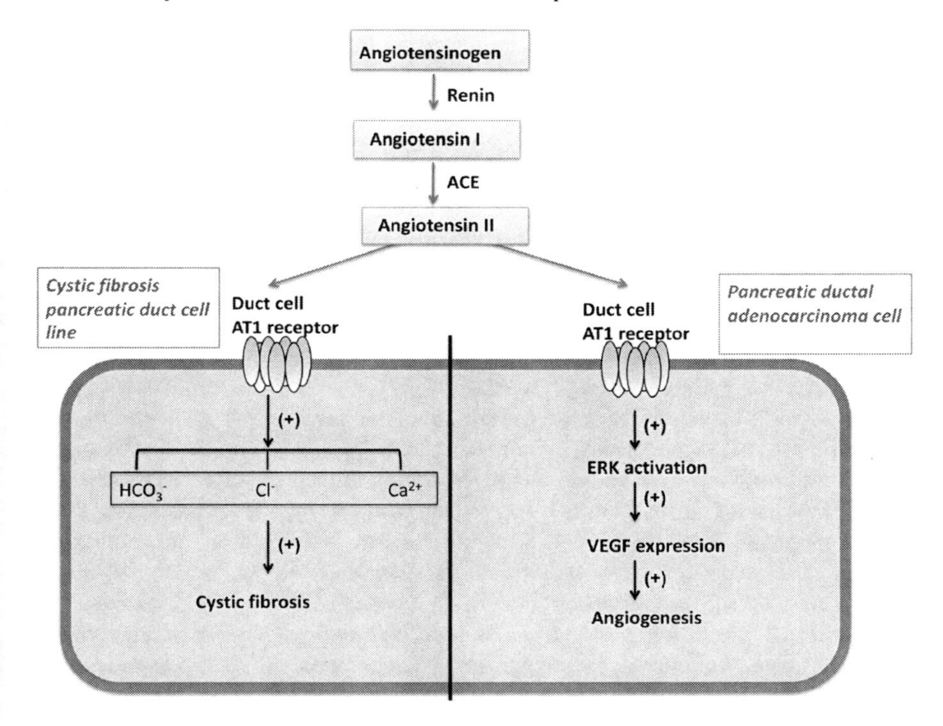

Fig. 6.2 A proposed existence and functional implications of a ductal cell RAS in the pancreas

6.3 Stellate Cell RAS

Like hepatic stellate cells, pancreatic stellate cells (PSCs) are vitamin A-storing cells and they contribute to fibrosis, a condition characterized by an excess deposition of extracellular matrix proteins (Kordes et al., 2009). PSCs are located at the base of pancreatic acini. They exhibit a radial pattern of cytoplasmic extension, encircling the basal aspect of acinar cells (Omary et al., 2007). In healthy pancreas, PSCs exist in a quiescent state, characterized by the presence of vitamin A-containing lipid droplets in their cytoplasm. In diseased states and under stimulation of growth factors, cytokines and oxidative stress, PSCs undergo proliferation, become responsive to proliferative and profibrogenic factors, and transform into a myofibroblast-like phenotype in which they express α-smooth muscle actin (α-SMA) and synthesize excess amounts of extracellular matrix (ECM) proteins, such as collagen and fibronectin (Apte et al., 1998; Shimizu, 2008). In fact, they are the major source of the ECM that contributes to fibrous tissue during pancreatic injury, notably pancreatic fibrosis, chronic pancreatitis and pancreatic cancers, as evidenced by previous studies (Apte et al., 1999; Masamune et al., 2009). In view of this unique property of PSCs, it is crucial to investigate some novel candidates that may be able to regulate the conversion of quiescent PSCs into activated PSCs. This line of work may lead to the development of new treatments for the management of pancreatic inflammation (Omary et al., 2007). Among these potential candidates, the RAS in the PSCs is emerging as a key mediator which is involved in the regulation of pancreatic fibrosis. The expression and localization of RAS components in PSCs along with the potential role of a local RAS in pancreatic fibrogenesis are discussed henceforth.

Stellate cells in the liver, called hepatic stellate cells (HSCs), are known to play a critical role in liver fibrosis. Previous studies have shown that the RAS is a critical player of hepatic fibrogenesis during liver injury (Wynn, 2008). Activated HSCs were found to produce angiotensin II, thus inducing contraction and proliferation of HSCs (Bataller et al., 2000). Meanwhile, RAS inhibition through AT_1 receptor antagonism was shown to attenuate liver fibrosis in both in vivo and in vitro studies (Jonsson et al., 2001; Bataller et al., 2000). In addition, angiotensin II and AT_1 receptor in lung stellate (Wang et al., 1999) and kidney stellate cells (Kida et al., 2007) have also been implicated in lung fibrosis and renal fibrosis, respectively. In the pancreas, PSCs are subjected to regulation by various endocrine, paracrine and autocrine factors, in particular, the RAS. In this context, PSCs have local RAS components including the AT_1 receptor, the antagonism with losartan of which plays a role in the development of pancreatic fibrogenesis (Liu et al., 2005; Nagashio et al., 2004; Reinehr et al., 2004). In situ hybridization and immunocytochemical experiments have shown that expression of AT_1 receptor mRNA and protein are localized mainly to cell membrane in isolated human PSCs; however, no angiotensin II was detected in culture media or cell homogenate of isolated human PSCs using radioimmunoassay, suggesting that angiotensin II utilizes paracrine, but not autocrine, signalling pathway via AT_1 receptors in PSCs (Liu et al., 2005). mRNA expression of several RAS components (angiotensinogen, ACE, AT_{1a}, AT_{1b} and AT_2) was also detected in isolated rat PSCs (Hama et al., 2004). Meanwhile,

AT_1 receptor protein expression was confirmed in PSCs which are upregulated by exposure to high glucose concentrations (Ko et al., 2006); additionally, α-SMA- and AT_1 receptor-positive PSCs have been observed in mice (Nagashio et al., 2004).

Converging evidence of the presence of a stellate cell RAS indicates its clinical relevance to pancreatic fibrogenesis such as pancreatitis and pancreatic cancer. RAS activation has been shown to play important roles in various tissue fibrosis and inflammation processes in RAS blockade studies (Schieffer et al., 1994, Ishidoya et al., 1995, Yoshiji et al., 2001). Locally produced angiotensin II promotes cell inflammation, ECM protein secretion, and collagen degradation (Ko et al., 2006). Angiotensin II could enhance DNA synthesis in PSCs by over-expression of Smad7 via the protein kinase C pathway (Hama et al., 2006). Angiotensin II could also enhance DNA synthesis by activating ERK through EGF receptor trans-activation, thus triggering the proliferation of activated PSCs (Hama et al., 2006), a key event in pancreatic fibrosis. Thus, inappropriate PSC activation via angiotensin II-mediated signalling in chronic pancreatitis and pancreatic cancer represents a potential target for intervention, such as by RAS blockade (Omary et al., 2007). Indeed, RAS inhibition could, by influencing PSC activation and proliferation, attenuate pancreatic inflammation and fibrosis (Kuno et al., 2003; Yamada et al., 2003). In experiments using an animal model of T2DM (Otsuka Long Evans Tokushima fatty rats), markers of islet fibrosis were decreased following application of an ACE inhibitor, and α-SMA positive PSCs and pancreatic ECM production were significantly reduced following ramipril treatment (Yoshikawa et al., 2002; Ko et al., 2004). In isolated PSCs exposed to high glucose levels for 24 h, AT_{1a} receptor subtype and ACE mRNA expression were upregulated without any change in expression of AT_{1b} receptor subtype and AT_2 receptors while angiotensin II production was increased. Interestingly, ACE inhibition and AT_1 receptor antagonism improved ECM production and islet fibrosis in PSCs under high glucose exposure (Ko et al., 2006). Furthermore, AT_{1a} receptor knockout (−/−) mice were reported to have reduced PSC activation relative to wild-type mice. When pancreatic fibrosis was induced by repeated episodes of cerulean injection-induced acute pancreatitis, pancreatic fibrosis was attenuated in AT_{1a} receptor (−/−) mice compared with that in wild-type mice (Nagashio et al., 2004). In light of these findings, these data indicate that the AT_1 receptor plays an important role in the development of pancreatic fibrosis through PSC activation and proliferation. In summary, the potential functional relevance of the pancreatic stellate cell RAS to fibrosis is depicted in Fig. 6.3.

6.4 Islet Cell RAS

Only 1–2% of the mass of pancreatic cells are of endocrine origin, namely the Islets of Langerhans. Among the five cell types found in islets (see Chapter 1, Section 1.1), the β-cell is the major cell type. The β-cell produces the peptide hormone insulin in response to nutrients (the primary stimulus) and hormonal and neural regulation, and plays a critical role in the maintenance of glucose homeostasis. Other peptides and proteins have recently been suggested to have profound impacts on pancreatic

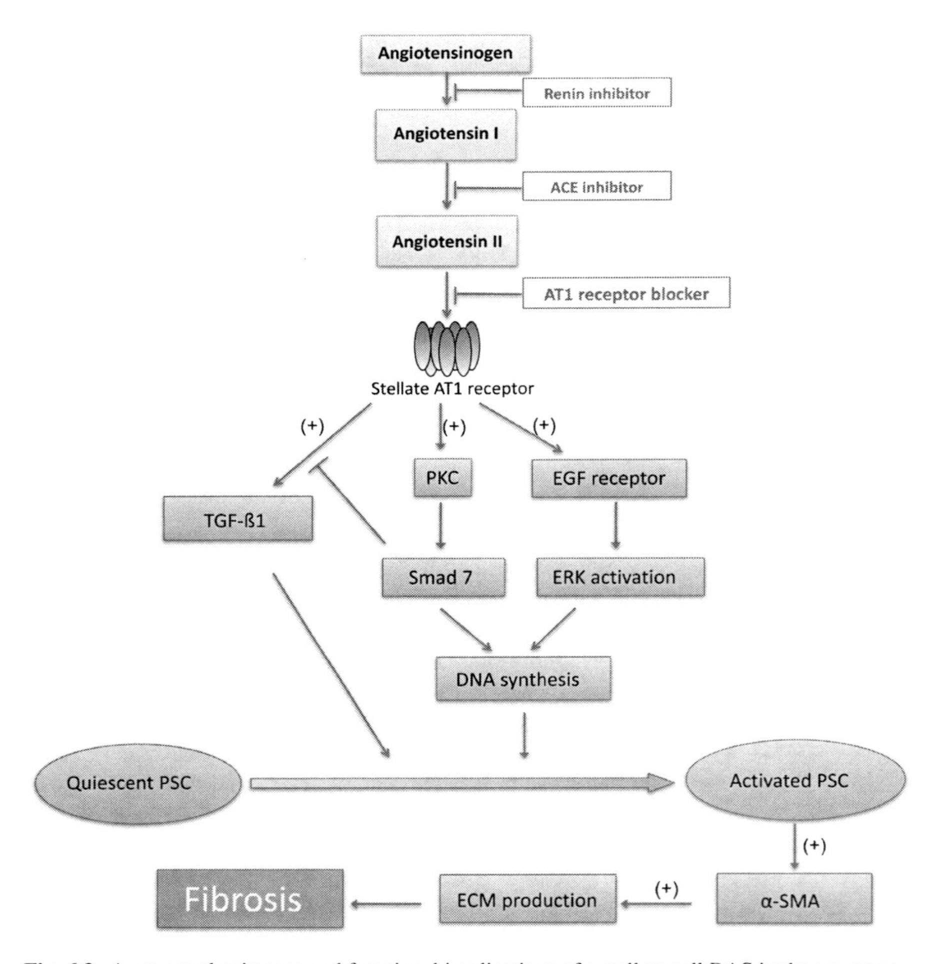

Fig. 6.3 A proposed existence and functional implications of a stellate cell RAS in the pancreas

islet secretion, differentiation, and growth as well as insulin sensitivity, and thus glucose metabolism and diabetes mellitus pathology (Leung & De Gasparo, 2009). Among these peptide candidates, the RAS has emerged as a potentially important regulator of islet physiology and pathophysiology (Leung, 2007). The expression, localization, regulation, and clinical relevance for T2DM of a pancreatic islet RAS have been critically reviewed in detail elsewhere (Leung & Carlsson, 2005; Leung, 2007; Chu & Leung, 2009). In the following section, solid evidence for the presence of several key RAS components in different islet cell types from the pancreas, and their potential roles in islet β-cell function and dysfunction from our laboratory and others, are discussed below.

(Pro)renin mRNA was previously localized to reticular fibers within pancreatic islets and endothelial cells of the pancreatic vasculature using in situ hybridization, while (pro)renin protein was localized predominantly to the β-cells themselves in human pancreas (Tahmasebi et al., 1999). In view of this observation, the authenticity of locally expressed (pro)renin within islets remained in question as (pro)renin could potentially be synthesized in the islet interstitium and taken up by the islets themself (Sealey et al., 1996; Tahmasebi et al., 1999). In term of angiotensinogen the mandatory component of a local RAS, researchers had been unable to detect its mRNA expression in rat pancreas for some time, perhaps due partly to technical limitations of mRNA assays, i.e. low detection sensitivity, as well as mRNA degradation during its extraction from the RNase-rich pancreas (Campbell & Habener, 1986). Recently, however, angiotensinogen mRNA was unequivocally identified in isolated mouse pancreatic islets using quantitative RT-PCR, confirming the existence of an islet RAS in the pancreas (Lau et al., 2004). Meanwhile angiotensinogen expression at both mRNA and protein levels in β-cells were shown to be markedly upregulated in transplanted islets and islets in diabetic mice (Lau et al., 2004; Chu et al., 2006). Interestingly, another recent study demonstrated expression and localization of angiotensinogen in islet α-cells of the rat pancreas (Regoli et al., 2003). In this study, immunoreactivity for angiotensinogen was detected at the periphery of the islets and most of the angiotensinogen-immunoreactive cells were predominantly glucagon-positive, a property consistent with α-cell identity. It is thus suggested that angiotensinogen and glucagon are secreted by different pathways in α-cells: glucagon is secreted through a regulated pathway, while angiotensinogen is produced through a constitutive pathway (Sernia, 1995; Regoli et al., 2003). Meanwhile, ACE and ACE2 mRNA and protein were also found in rat pancreatic β-cells, localized mainly to the microvasculature and the islet periphery. Concurrently, ACE and ACE2 expression in islets were also found to be upregulated in the Zucker diabetic fatty rat, a rat model of obesity-induced T2DM (Tikellis et al., 2004).

In the human pancreas, expression of both AT_1 and AT_2 receptors have been identified in pancreatic islets, with AT_1 receptors being located mainly in cells near the centre of the islets and co-localized with insulin secreting β-cells (Tahmasebi et al., 1999). In the mouse pancreas, AT_1 and AT_2 receptor transcription and translation products were found in isolated mouse pancreatic islets; immunocytochemical data further confirmed that AT_1 receptor immunoreactivity was co-localized with that for insulin in islet β-cells (Lau et al., 2004). In addition, AT_1 receptor protein formation was found to be increased in transplanted islets (Lau et al., 2004) and islets from obesity-induced db/db mice of T2DM (Chu et al., 2006). In contrast, AT_2 receptor expression was predominantly co-localized with somatostatin secreting δ-cells in the periphery of rat islets (Wong et al., 2004). In keeping with these findings in rat pancreas, abundant expression of AT_2 receptors was also detected in two pancreatic endocrine cell lines, namely RIN-m and RIN-14B, which may be involved in the regulation of somatostatin release (Wong et al., 2004). In another pancreatic β-cell line, called INS-1E, AT_1 receptor immunoreactivity was found to be localized specifically on the cell membrane; when the INS-1E cells were exposed to high glucose concentrations, it resulted in dose-dependent upregulation of AT_1 receptor

Table 6.2 The expression and localization of key RAS components from different cell types of pancreatic islets

RAS components	Cell types in pancreatic islets	References
Renin/pro-renin	β cells and Islet vasculature	Tahmasebi et al. (1999) and Tikellis et al. (2004)
Angiotensinogen	α cells and β cells	Lau et al. (2004), Regoli et al. (2003) and Chu et al. (2006)
ACE	Islet microvasculature	Lau et al. (2004) and Chu et al. (2006)
ACE2	β cells and islet icrovasculature	Tikellis et al. (2004)
AT_1 receptor	β cells	Tahmasebi et al. (1999), Lau et al. (2004), Ko et al. (2004) and Chu et al. (2006)
AT_2 receptor	δ cells	Wong et al. (2004)
	β cells	Chu et al. (2010)

mRNA and protein expression (Leung & Leung, 2008). Taken together, key localization and expression studies of major RAS components in different cell types of the pancreatic islets are summarized in Table 6.2.

The aforementioned findings provide substantial evidence in favour of the existence of an islet RAS in the endocrine pancreas. The immediate question that needs to be raised is: what are the functional and clinical implications of an islet RAS? In this context, islet blood flow, which plays a key role in controlling islet secretion, is subjected to regulation by islet RAS activity. Angiotensin II, a potent vasoconstrictor, was demonstrated to reduce islet blood flow while RAS blockade with either ACE inhibitor or AT_1 receptor blocker, enhanced islet blood flow in rat perfusion studies (Carlsson et al., 1998). Consistently, the first phase of glucose-stimulated insulin release was impaired in animals subjected to acute angiotensin II treatment, and this effect was reversed by administration of an ACE inhibitor, an effect related to islet blood flow (Pollare et al., 1989). Furthermore, subsequent studies have shown that AT_1 receptor blockade can enhance islet blood and insulin secretion, thus improving glucose tolerance and glycemic control in rat (Huang et al., 2007). Besides being a potent vasoconstrictor on the islet blood flow thus insulin release, angiotensin II can also influence directly islet cell secretory function which is independent of its pancreatic blood flow. To address this issue, isolated islets devoid of vasculature were employed in our laboratory. Intriguingly, our results showed that angiotensin II dose-dependently inhibited insulin release in response to glucose stimulation, an effect that was completely rescued by AT_1 receptor antagonism (with losartan) in mouse isolated pancreatic islets. The regulatory mechanism by which angiotensin II induces inhibitory effects on glucose-stimulated insulin secretion may be due, at least in part, to decreased (pro)insulin biosynthesis rather than an effect on glucose oxidation rate (Lau et al., 2004). On the other hand, losartan selectively improved glucose-stimulated insulin secretion and (pro)insulin biosynthesis, and thus reduced hyperglycemia and glucose intolerance using a mouse model of T2DM

(Chu et al., 2006). Furthermore, the up-regulated NAD(P)H oxidase expression and activation were also mediated by islet AT_1 receptor (Hirata et al., 2009) while specific antagonism was shown to be beneficial for islet cell secretion and β-cell mass, probably as a result of effects on oxidative stress-induced islet apoptosis and proliferation (Chu & Leung, 2007; Cheng et al., 2008), AT_1 receptor-ERK1/2 dependent mechanism was potentially involved in the angiotensin II induced islet MCP-1 up-regulation, the co-localization of islet local ACE and MCP-1 indicated the endogenous interaction between islet RAS and islet inflammation (Chipitsyna et al., 2007). Taken these data together, all available findings indicate that islet cell secretory function and cell mass may be subjected to regulation by the RAS involving primarily the AT_1 receptor via three potential mechanistic pathways: islet blood flow, (pro)insulin biosynthesis, as well as oxidative stress and inflammation (Fig. 6.4). Further discussion on the research progress on the RAS in the endocrine pancreas concerning T2DM will be critically reviewed in Chapter 8.

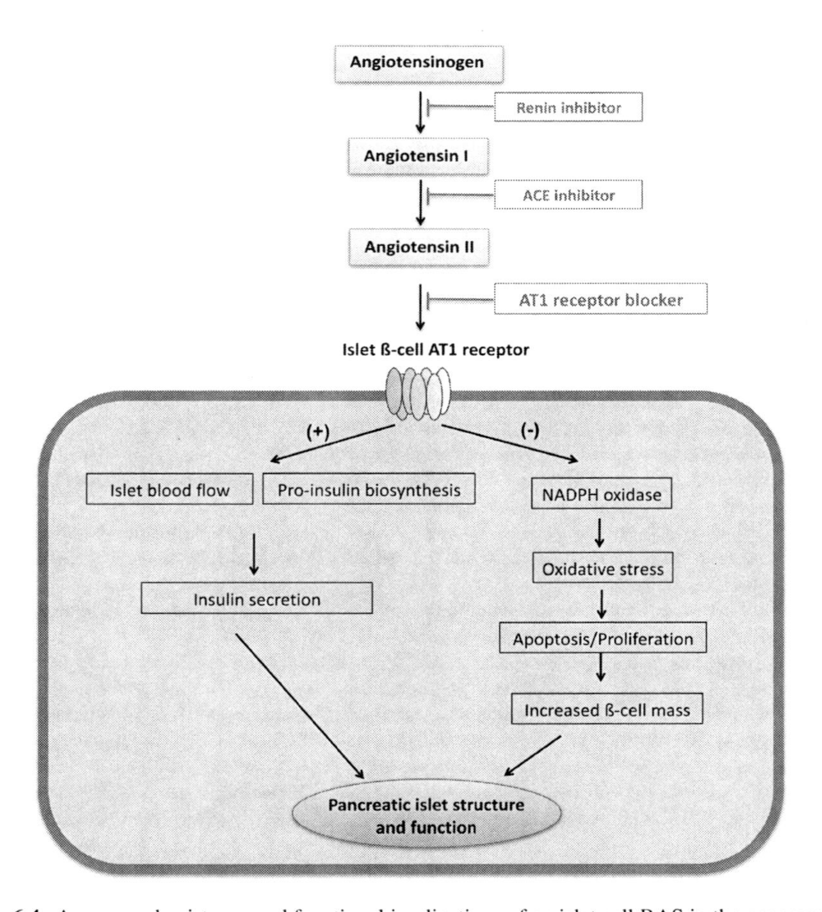

Fig. 6.4 A proposed existence and functional implications of an islet cell RAS in the pancreas

References

Abuissa H, Jones PG, Marso SP, O'Keefe JH Jr. Angiotensin-converting enzyme inhibitors or angiotensin receptor blockers for prevention of type 2 diabetes: a meta-analysis of randomized clinical trials. *J Am Coll Cardiol* **46**:821–826, 2005.

Anandanadesan R, Gong Q, Chipitsyna G, Witkiewicz A, Yeo CJ and Arafat HA. Angiotensin II induces vascular endothelial growth factor in pancreatic cancer cells through an angiotensin II type 1 receptor and ERK1/2 signaling. *J Gastrointest Surg* **12**:57–66, 2008.

Apte MV, Haber PS, Applegate TL, Norton ID, McCaughan GW, Korsten MA, Pirola RC and Wilson JS. Periacinar stellate shaped cells in rat pancreas: identification, isolation, and culture. *Gut* **43**:128–133, 1998.

Apte MV, Haber PS, Darby SJ, Rodgers SC, McCaughan GW, Korsten MA, Pirola RC and Wilson JS. Pancreatic stellate cells are activated by proinflammatory cytokines: implications for pancreatic fibrogenesis. *Gut* **44**:534–541, 1999.

Arafat HA, Gong Q, Chipitsyna G, Rizvi A, Saa CT and Yeo CJ. Antihypertensives as novel antineoplastics: angiotensin-I-converting enzyme inhibitors and angiotensin II type 1 receptor blockers in pancreatic ductal adenocarcinoma. *J Am Coll Surg* **204**:996–1005, 2007.

Bataller R, Gines P, Nicolas JM, Gorbig MN, Garcia-Ramallo E, Gasull X, Bosch J, Arroyo V and Rodes J. Angiotensin II induces contraction and proliferation of human hepatic stellate cells. *Gastroenterology* **118**:1149–1156, 2000.

Bateman A, Turner S, Thomas K, McCrudden P, Fine D, Johnson P, Johnson C and Iredale J. apoptosis and proliferation of acinar and islet cells in chronic pancreatitis: evidence for differential cell loss mediating preservation of islet function. *Gut* **50**:542–548, 2002.

Campbell D and Habener J. Angiotensinogen gene is expressed and differentially regulated in multiple tissues of the rat. *J Clin Invest* **78**:31–39, 1986.

Carlsson PO, Berne C and Jansson L. Angiotensin II and the endocrine pancreas: effects on islet blood flow and insulin secretion in rats. *Diabetologia* **41**:127–133, 1998.

Chan H, Law S, Leung P, Fu L and Wong P. Angiotensin II receptor type I-regulated anion secretion in cystic fibrosis pancreatic duct cells. *J Membr Biol* **156**:241–249, 1997.

Chappell M, Diz D and Gallagher P. The renin-angiotensin system and the exocrine pancreas. *JOP* **2**:33–39, 2001.

Chappell M, Jacobsen D and Tallant E. Characterization of angiotensin II receptor subtypes in pancreatic acinar AR42J cells. *Peptides* **16**:741–747, 1995.

Chappell M, Millsted A, Diz D, Brosnihan K and Ferrario C. Evidence for an intrinsic angiotensin system in the canine pancreas. *J Hypertens* **9**:751–759, 1991.

Chappell MC, Diz DI and Jacobsen DW. Pharmacological characterization of angiotensin II binding sites in the canine pancreas. *Peptides* **13**:313–318, 1992.

Chehl N, Gong Q, Chipitsyna G, Aziz T, Yeo CJ and Arafat HA. Angiotensin II Regulates the Expression of Monocyte Chemoattractant Protein-1 in Pancreatic Cancer Cells. *J Gastrointest Surg* **13**:2189–2200, 2009.

Cheng H, So S, Law S and Chan H. Angiotensin II-mediated signal transduction events in cystic fibrosis pancreatic duct cells. *Biochim Biophys Acta* **1449**:254–260, 1999.

Cheng Q, Law PK, de Gasparo M and Leung PS. Combination of the dipeptidyl peptidase IV inhibitor LAF237 [(S)-1-[(3-hydroxy-1-adamantyl)ammo]acetyl-2-cyanopyrrolidine] with the angiotensin II type 1 receptor antagonist valsartan [N-(1-oxopentyl)-N-[[2′-(1H-tetrazol-5-yl)-[1,1′-biphenyl]-4-yl]methyl]-L-valine] enhances pancreatic islet morphology and function in a mouse model of type 2 diabetes. *J Pharmacol Exp Ther* **327**:683–691, 2008.

Cheung WT, Yeung SY, Yiu AK, Ip TM, Wan DC, Luk SK and Ho WK. Characterization of a functional AT1A angiotensin receptor in pancreatoma AR4-2 J cells. *Peptides* **20**:829–836, 1999.

Chipitsyna G, Gong Q, Gray CF, Haroon Y, Kamer E, Arafat HA (2007) Induction of monocyte chemoattractant protein-1 expression by angiotensin II in the pancreatic islets and beta-cells. *Endocrinology* **148**:2198–2208, 2007.

Christophe J. Pancreatic tumoral cell line AR42J: an amphicrine model. *Am J Physiol* **266**: G963–G971, 1994.

Chu KY, Lau T, Carlsson PO and Leung PS. Angiotensin II type 1 receptor blockade improves beta-cell function and glucose tolerance in a mouse model of type 2 diabetes. *Diabetes* **55**:367–374, 2006.

Chu KY and Leung PS. Angiotensin II Type 1 receptor antagonism mediates uncoupling protein 2-driven oxidative stress and ameliorates pancreatic islet beta-cell function in young Type 2 diabetic mice. *Antioxid Redox Signal* **9**:869–878, 2007.

Chu KY and Leung PS. Angiotensin II in type 2 diabetes mellitus. *Curr Protein Pept Sci* **10**:75–84, 2009.

Chu KY, Cheng Q, Chen C, Au LS, Seto SW, Tuo Y, Motin L, Kwan YW and Leung PS. Angiotensin II exerts glucose-dependent effects on Kv currents in mouse pancreatic beta-cells via angiotensin II type 2 receptors. *Am J Physiol Cell Physiol* **298**:C313–323, 2010.

Cox HM, Cuthbert AW and Munday KA. The effect of angiotensin II upon electrogenic ion transport in rat intestinal epithelia. *Br J Pharmacol* **90**:393–401, 1987.

Fink AS, Wang Y, Mendez T, Worrell RT, Eaton D, Nguyen TD and Lee SP. Angiotensin II evokes calcium-mediated signaling events in isolated dog pancreatic epithelial cells. *Pancreas* **25**: 290–295, 2002.

Fujita M, Hayashi I, Yamashina S, Itoman M and Majima M. Blockade of angiotensin AT1a receptor signaling reduces tumor growth, angiogenesis, and metastasis. *Biochem Biophys Res Commun* **294**:441–447, 2002.

Garber AJ. Perspectives in type 2 diabetes: incorporating the latest insulin analogue strategies to achieve treatment success. *Diabetes Obes Metab* **11**(Suppl 5):III–IV, 2009.

Githens S. The pancreatic duct cell: proliferative capabilities, specific characteristics, metaplasia, isolation, and culture. *J Pediatr Gastroenterol Nutr* **7**:486–506, 1988.

Hama K, Ohnishi H, Aoki H, Kita H, Yamamoto H, Osawa H, Sato K, Tamada K, Mashima H, Yasuda H and Sugano K. Angiotensin II promotes the proliferation of activated pancreatic stellate cells by Smad7 induction through a protein kinase C pathway. *Biochem Biophys Res Commun* **340**:742–750, 2006.

Hama K, Ohnishi H, Yasuda H, Ueda N, Mashima H, Satoh Y, Hanatsuka K, Kita H, Ohashi A, Tamada K and Sugano K. Angiotensin II stimulates DNA synthesis of rat pancreatic stellate cells by activating ERK through EGF receptor transactivation. *Biochem Biophys Res Commun* **315**:905–911, 2004.

Hashimoto T, Yamada T, Yokoi T, Sano H, Ando H, Nakazawa T, Ohara H, Nomura T, Joh T and Itoh M. Apoptosis of acinar cells is involved in chronic pancreatitis in Wbn/Kob rats: role of glucocorticoids. *Pancreas* **21**:296–304, 2000.

Hirata AE, Morgan D, Oliveira-Emilio HR, Rocha MS, Carvalho CR, Curi R and Carpinelli AR. Angiotensin II induces superoxide generation via NAD(P)H oxidase activation in isolated rat pancreatic islets. *Regul Pept* **153**:1–6, 2009.

Howard-McNatt M and Fink AS. Captopril inhibits secretin-induced pancreatic bicarbonate output. *J Surg Res* **103**:96–99, 2002.

Huang Z, Jansson L and Sjoholm A. Vasoactive drugs enhance pancreatic islet blood flow, augment insulin secretion and improve glucose tolerance in female rats. *Clin Sci (Lond)* **112**: 69–76, 2007.

Ishidoya S, Morrissey J, McCracken R, Reyes A and Klahr S. Angiotensin II receptor antagonist ameliorates renal tubulointerstitial fibrosis caused by unilateral ureteral obstruction. *Kidney Int* **47**:1285–1294, 1995.

Jonsson JR, Clouston AD, Ando Y, Kelemen LI, Horn MJ, Adamson MD, Purdie DM and Powell EE. Angiotensin-converting enzyme inhibition attenuates the progression of rat hepatic fibrosis. *Gastroenterology* **121**:148–155, 2001.

Kida Y, Asahina K, Inoue K, Kawada N, Yoshizato K, Wake K and Sato T. Characterization of vitamin A-storing cells in mouse fibrous kidneys using Cygb/STAP as a marker of activated stellate cells. *Arch Histol Cytol* **70**:95–106, 2007.

Ko S, Kwon H, Kim S, Moon S, Ahn Y, Song K, Son H, Cha B, Lee K, Son H, Kang S, Park C, Lee I and Yoon K. Ramipril treatment suppresses islet fibrosis in Otsuka Long–Evans Tokushima fatty rats. *Biochem Biophys Res Commun* **316**:114–122, 2004.

Ko SH, Hong OK, Kim JW, Ahn YB, Song KH, Cha BY, Son HY, Kim MJ, Jeong IK and Yoon KH. High glucose increases extracellular matrix production in pancreatic stellate cells by activating the renin-angiotensin system. *J Cell Biochem* **98**:343–355, 2006.

Korc M. Pathways for aberrant angiogenesis in pancreatic cancer. *Mol Cancer* **2**:1–8, 2003.

Kordes C, Sawitza I and Haussinger D. Hepatic and pancreatic stellate cells in focus. *Biol Chem* **390**:1003–1012, 2009.

Kuno A, Yamada T, Masuda K, Ogawa K, Sogawa M, Nakamura S, Nakazawa T, Ohara H, Nomura T, Joh T, Shirai T and Itoh M. Angiotensin-converting enzyme inhibitor attenuates pancreatic inflammation and fibrosis in male Wistar Bonn/Kobori rats. *Gastroenterology* **124**:1010–1019, 2003.

Lam KY and Leung PS. Regulation and expression of a renin-angiotensin system in human pancreas and pancreatic endocrine tumours. *Eur J Endocrinol* **146**:567–572, 2002.

Lau T, Carlsson P and Leung P. Evidence for a local angiotensin-generating system and dose-dependent inhibition of glucose-stimulated insulin release by angiotensin II in isolated pancreatic islets. *Diabetologia* **47**:240–248, 2004.

Leung PS and De Gasparo M. Novel peptides and proteins in diabetes mellitus. *Curr Protein Pept Sci* **10**:1, 2009.

Leung KK and Leung PS. Effects of hyperglycemia on angiotensin II receptor type 1 expression and insulin secretion in an INS-1E pancreatic beta-cell line. *JOP* **9**:290–299, 2008.

Leung PS. The physiology of a local renin-angiotensin system in the pancreas. *J Physiol* **580**: 31–37, 2007.

Leung PS and Carlsson PO. Pancreatic islet renin-angiotensin system: its novel roles in islet function and in diabetes. *Pancreas* **30**:293–298, 2005.

Leung PS and Chappell MC. A local pancreatic renin-angiotensin system: endocrine and endocrine roles. *Int J Biochem Cell Biol* **35**:838–846, 2003.

Leung PS, Chan WP, Wong TP and Sernia C. Expression and localization of the renin-angiotensin system in the rat pancreas. *J Endocrinol* **160**:13–19, 1999.

Leung PS, Chan HC and Wong PY. Immunohistochemical localization of angiotensin II in the mouse pancreas. *Histochem J* **30**:21–25, 1998.

Leung PS, Chan HC, Fu LX and Wong PY. Localization of angiotensin II receptor subtypes AT1 and AT2 in the pancreas of rodents. *J Endocrinol* **153**:269–274, 1997a.

Leung PS, Chan HC, Fu LX, Zhou WL and Wong PY. Angiotensin II receptors, AT1 and AT2 in the rat epididymis: immunocytochemical and electrophysiological studies. *Biochim Biophys Acta* **1357**:65–72, 1997b.

Liu WB, Wang XP, Wu K and Zhang RL. Effects of angiotensin II receptor antagonist, Losartan on the apoptosis, proliferation and migration of the human pancreatic stellate cells. *World J Gastroenterol* **11**:6489–6494, 2005.

Masamune A, Watanabe T, Kikuta K and Shimosegawa T. Roles of pancreatic stellate cells in pancreatic inflammation and fibrosis. *Clin Gastroenterol Hepatol* **7**:S48–S54, 2009.

Musch MW, Li YC and Chang EB. Angiotensin II directly regulates intestinal epithelial NHE3 in Caco2BBE cells. *BMC Physiology* **9**:1–8, 2009.

Nagashio Y, Asaumi H, Watanabe S, Nomiyama Y, Taguchi M, Tashiro M, Sugaya T and Otsuki M. Angiotensin II type 1 receptor interaction is an important regulator for the development of pancreatic fibrosis in mice. *Am J Physiol* **287**:G170–G177, 2004.

Norris B, Gonzalez C, Concha J, Palacios S and Contreras G. Stimulatory effect of angiotensin II on electrolyte transport in canine tracheal epithelium. *Gen Pharmacol* **22**:527–531, 1991.

Omary MB, Lugea A, Lowe AW and Pandol SJ. The pancreatic stellate cell: a star on the rise in pancreatic diseases. *J Clin Invest* **117**:50–59, 2007.

Paul M, Poyan Mehr A and Kreutz R. Physiology of local renin-angiotensin systems. *Physiol Rev* **86**:747–803, 2006.

Pollare T, Lithell H and Berne C. A comparison of the effects of hydrochlorothiazide and captopril on glucose and lipid metabolism in patients with hypertension. *N Engl J Med* **321**:868–873, 1989.

Regoli M, Bendayan M, Fonzi L, Sernia C and Bertelli E. Angiotensinogen localization and secretion in the rat pancreas. *J Endocrinol* **179**:81–89, 2003.

Reinehr R, Zoller S, Klonowski-Stumpe H, Kordes C and Haussinger D. Effects of angiotensin II on rat pancreatic stellate cells. *Pancreas* **28**:129–137, 2004.

Schieffer B, Wirger A, Meybrunn M, Seitz S, Holtz J, Riede UN and Drexler H. Comparative effects of chronic angiotensin-converting enzyme inhibition and angiotensin II type 1 receptor blockade on cardiac remodeling after myocardial infarction in the rat. *Circulation* **89**: 2273–2282, 1994.

Sealey JE, Catanzaro DF, Lavin TN, Gahnem F, Pitarresi T, Hu LF and Laragh JH. Specific prorenin/renin binding (ProBP). Identification and characterization of a novel membrane site. *Am J Hypertens* **9**:491–502, 1996.

Sernia C. Location and secretion of brain angiotensinogen. *Regul Pept* **57**:1–18, 1995.

Shimizu K. Mechanisms of pancreatic fibrosis and applications to the treatment of chronic pancreatitis. *J Gastroenterol* **43**:823–832, 2008.

Tahmasebi M, Puddefoot J, Inwang E and Vinson G. The tissue renin-angiotensin system in human pancreas. *J Endocrinol* **161**:317–322, 1999.

Tikellis C, Wookey P, Candido R, Andrikopoulos S, Thomas M and Cooper M. Improved islet morphology after blockade of the renin-angiotensin system in the ZDF rat. *Diabetes* **53**: 989–997, 2004.

Tsang SW, Cheng HK, Leung PS. The role of the pancreatic renin-angiotensin system in acinar digestive enzyme secretion and in acute pancreatitis. *Regul Pept* **119**:213–219, 2004.

Wang R, Ramos C, Joshi I, Zagariya A, Pardo A, Selman M and Uhal BD. Human lung myofibroblast-derived inducers of alveolar epithelial apoptosis identified as angiotensin peptides. *Am J Physiol* **277**:L1158–L1164, 1999.

Wong P, Lee S and Cheung W. Immunohistochemical colocalization of type II angiotensin receptors with somatostatin in rat pancreas. *Regul Pept* **117**:195–205, 2004.

Wynn TA. Cellular and molecular mechanisms of fibrosis. *J Pathol* **214**:199–210, 2008.

Yamada T, Kuno A, Masuda K, Ogawa K, Sogawa M, Nakamura S, Ando T, Sano H, Nakazawa T, Ohara H, Nomura T, Joh T and Itoh M. Candesartan, an angiotensin II receptor antagonist, suppresses pancreatic inflammation and fibrosis in rats. *J Pharmacol Exp Ther* **307**:17–23, 2003.

Yin G, Yan C and Berk BC. Angiotensin II signaling pathways mediated by tyrosine kinases. *Int J Biochem Cell Biol* **35**:780–783, 2003.

Yoshiji H, Kuriyama S, Yoshii J, Ikenaka Y, Noguchi R, Nakatani T, Tsujinoue H and Fukui H. Angiotensin-II type 1 receptor interaction is a major regulator for liver fibrosis development in rats. *Hepatology* **34**:745–750, 2001.

Yoshikawa H, Kihara Y, Taguchi M, Yamaguchi T, Nakamura H and Otsuki M. Role of TGF-beta1 in the development of pancreatic fibrosis in Otsuka Long–Evans Tokushima Fatty rats. *Am J Physiol Gastrointest Liver Physiol* **282**:G549–G558, 2002.

Part III
Research Progress of the RAS in Pancreas

Chapter 7
Basic Techniques for Pancreatic Research

7.1 Cell Models

Cell culture system studies complement animal studies, providing an indispensable tool for examining the underlying signalling mechanisms by which certain physiological responses are evoked and, at the same time, allowing observation to be attributed to distinct individual treatment factor. They are often applied as primary screenings of the potential effects of certain drug treatment prior to further animal studies. Experiments utilizing pancreatic cell lines or primary pancreatic cell cultures have been widely reported. The term cell line generally refers to an immortalized group of cells originally derived from a single clone which can be, in principle, maintained permanently and proliferated upon exposure to appropriate medium in vitro. Many cell lines are derived from abnormal tissues, such as cancer specimens or tumour cells.

The pancreatic cell lines that are most commonly used in experiments, along with some pertinent information and relevant literature for each cell line model, are discussed and summarized in this Chapter (see Table 7.1). Pancreatic cancer cell lines are divided into exocrine (adenocarcinoma) and endocrine tumor classes. It is worth noting that a primary culture is different from a pure cell line; primary culture refers to cells or tissues that are taken directly from the source tissue specimens. Primary cultures generally include populations of cells. Protocols for the isolation and culture of individual cell type components (e.g., acinar and islet cells) from the pancreas are well established in the field and are discussed briefly in this chapter.

An important advance in cell culture is the development of the co-culture system. Under certain conditions, different cells are grown together in a mixture; and in some protocols, the cells are separated while allowing the flow of secretory molecules through a permeable membrane insert. The choice for the membrane inserts with an appropriate pore size is critical to best suit for different experimental needs. A large pore size allows not only the exchange of secretory molecules but also the migration of cultured cells between the two compartments. In the context of pancreatic cancer research, such application becomes a useful technique in studying the metastatic properties of the cells under different microenvironment (Chen et al., 2009). Local invasion of cells can often be directly captured using

P.S. Leung, *The Renin-Angiotensin System: Current Research Progress in The Pancreas*, Advances in Experimental Medicine and Biology 690,
DOI 10.1007/978-90-481-9060-7_7, © Springer Science+Business Media B.V. 2010

Table 7.1 The common cell lines used in basic pancreatic research

Disease of origin		Species	Names	References
Exocrine tumor (Adenocarcinoma)	Acinar cell tumor	Mouse	266-6	Siveke et al. (2008)
		Rat	AR42J	Chan & Leung (2009)
	Ductal/epithelial cell tumor	Mouse	LTPA	Leiter et al. (1978)
		Rat	ARIP	Wang et al. (2001)
		Human	PL45	Suemizu et al. (2007)
			Capan-2	Shi et al. (2008)
			PANC-1	Lau et al. (2008)
			Capan-1[a]	
			SW-1990[b]	
Endocrine tumor	Glucagonoma	Mouse	αTC1	Chuang et al. (2008)
		Rat	AN 697	Petersen et al. (2000)
	Insulinoma	Mouse	β-TC-6	Mwangi et al. (2008)
			MIN6	Kaneko et al. (2009)
		Rat	RIN-m5F	Onoue et al. (2008)
			INS-1	Leung & Leung (2008)
		Hamster	HIT-T15	Mariogo et al. (2009)
		Human	NES2Y	Ou et al. (2005)

[a] Cells derived from liver metastasis site.
[b] Cells derived from spleen metastasis site.

conventional microscopes (Farrow et al., 2009). The use of a co-culture system reveals the importance of cell-to-cell interactions. In this context, a recent report elegantly revealed, using a rotational co-culture system that the pancreatic duct-derived epithelial cells assist in maintaining the structural integrity and functional viability of isolated human islets (Murray et al., 2009). Co-culture paradigms are also commonly utilized in pancreatic stem cell research wherein the stem cells are grown and developed under the influence of specified factors secreted by another cell type within the co-culture system. One prominent application in the field is exemplified by the maintenance of embryonic stem cells on a feeder cell layer (Zhou et al., 2008).

Apart from culturing pancreatic cell lines or isolated primary cell culture, the whole pancreas can also be isolated and cultured under suitable conditions (Parsa & Marsh, 1976). This organ culture technique is commonly seen in studying embryonic pancreas which serves as a good platform to reveal its developmental changes. A suitable culturing medium supplemented with enough growth factors is required so that the pancreas will be developed to mimic the actual developmental scenario in the embryos. Typically, the dissected pancreatic rudiments from rodent and/or human embryos are laid on the permeable membrane inserts at the air-liquid interface and the culture can be sustained for about one week (Guillemain et al., 2007; Castaing et al., 2005). Some studies have also reported with the use of the hanging drop technique when culturing the embryonic pancreas (Tei et al., 2005).

7.2 Animal Models

The use of animal models is very useful in the development of novel therapeutic strategies and therefore the creation of experimental models is of major emphasis on our current research. Various animal models are available to represent the clinically relevant conditions of Type 1 and Type 2 diabetes mellitus (T1DM and T2DM, respectively). Figure 7.1 and Table 7.2 provide a summary of rodent models of T1DM and T2DM which are commonly employed in contemporary research on diabetes.

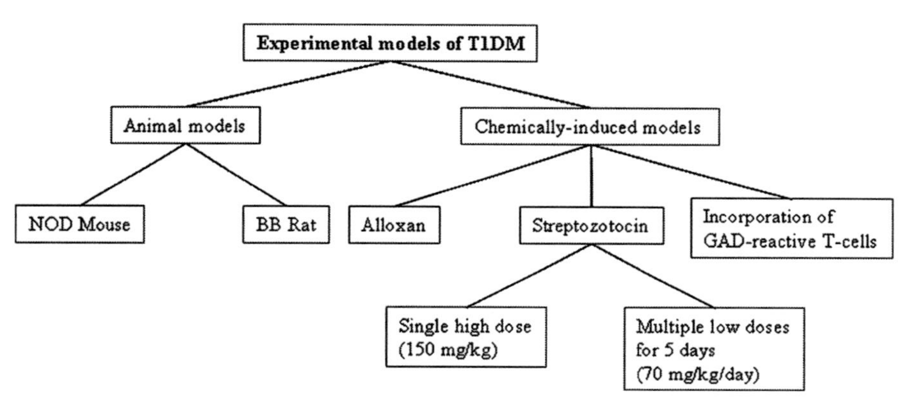

Fig. 7.1 A summary of the common experimental models of T1DM in basic pancreatic research. NOD, Non-obese diabetic; BB, Bio-breeding; GAD, glutamic acid decarboxylase

Table 7.2 The common rodent models of T2DM and their characteristics in basic pancreatic research (the table is modified from Masiello, 2006, Kadowaki, 2000)

		Obesity	Associated insulin resistance	Phenotype
Knockout mouse models	db/db	Yes	Yes	Severe diabetes at adult age
	PDX-1 +/–	No	No	Impaired glucose tolerance
	IRS-2 +/–	No	Yes	Mild to severe diabetes
	GLUT4 –/–	No	Yes	Not diabetic
Rat models	ZDF rat	Yes	Yes	Mild diabetes at adult age
	GK rat	No	No	Mild diabetes at adult age
	n-STZ rat	No	No	Mild hyperglycemia

PDX-1, pancreatic duodenal homeobox factor 1; IRS-2, insulin receptor substrate-2; Glut4, solute carrier family 2 (facilitated glucose transporter), member 4; ZDF, zucker diabetic fatty; GK, Goto-Kakizaki; n-STZ, netonatal streptozotocin.

7.2.1 Animal Models of T1DM

Typical T1DM animal models include Non-obese Diabetic (NOD) mice and Bio-breeding (BB) rats. The genetic backgrounds of both rodents were designed specifically to contain many genes related to susceptibility to autoimmunity, thus producing spontaneous induction of T1DM in the adult animals (von Herrath & Nepom, 2009). These experimental T1DM models provide a good platform for basic scientists and clinicians so as to facilitate research aimed at elucidating the mechanistic pathways involved in diabetes onset and thus informing potential avenues of prevention. The presence of some specific cytokines was recently demonstrated to be necessary for the development of T1DM (Sutherland et al., 2009). Only subtle differences in T1DM onset have been identified between NOD mouse and BB rat studies. Briefly, there appears to be no sex difference in T1DM onset in BB rats (Verheul et al., 1986); meanwhile female NOD mice often become permanently hyperglycemic at an earlier age and at a higher frequency than male NOD mice (Atkinson & Leiter, 1999; Leiter et al., 1987).

Chemical induction strategies are also commonly employed to generate animal models of T1DM. Early studies used alloxan, a glucose analogue that is toxic to β-cells, to produce experimental diabetes (Lenzen & Panten, 1988). More recently, streptozotocin (STZ) has been purported as a replacement chemical for alloxan. STZ has been widely utilized to induce T1DM because of its greater selectivity for β-cells and longer systemic half-life (Junod et al., 1969). STZ is a naturally occurring chemical that is specifically destructive to the insulin-producing β-cells in vivo. It is structurally similar to glucose, and thus can be transported into the β-cells through GLUT-2, but not other types of glucose transporters (Schnedl et al., 1994). There are two major methods for STZ-induced diabetes in rodents, including a single high dose administration (150 mg/kg) (Suen et al., 2006) and a multiple low dose administration (70 mg/kg/day) for consecutive 5 days (Kroon et al., 2008). Experimental findings have shown subtle differences in the dynamics of diabetes formation by these two modes of injection although there is a general observable phenomenon that animals receiving a high-dose STZ administration do not display a higher mortality rate. A multiple low-dose administration method has also been reported to generate a better animal model of possessing progression and long-term complications of diabetes (Arora et al., 2009). Apart from the typical use of STZ to generate T1DM animal models, other study has reported the incorporation of glutamic acid decarboxylase (GAD)-reactive T-cells into NOD mice so as to trigger the attack of the host's islet cells and thus accelerates the onset of diabetes (Zekzer et al., 1998).

7.2.2 Animal Models of T2DM

Various T2DM animal models which mimic the disease are also well established (LeRoith & Gavrilova, 2006; Cefalu, 2006). A key clinical characteristic of patients with T2DM is the presence of peripheral insulin resistance. Thus knockout mice have been bred to develop insulin resistance and T2DM. Numerous knockout mice models of T2DM have been developed including, but name a few, knockout of genes for the following proteins: insulin receptor substrate-2, GLUT-4, phosphoinositide-3 kinase, and peroxisome proliferator–activated receptor γ. Knockout of such genes interferes with normal insulin signalling, thus influencing glucose homeostasis or the normal cellular glucose metabolism (Kadowaki, 2000; Masiello, 2006; Brissova et al., 2002; Katz et al., 1995). A number of transgenic T2DM models are involved in the knockout of phenotypic factors along the pancreatic development. Pancreatic duodenal homeobox factor 1 (PDX-1)-knockout mice is one of the common seen T2DM animal models which falls into this category. A heterozygous mutation of the PDX-1 gene is linked with one of the commonly known maturity onset diabetes of the young (MODY). As PDX-1 is a critical transcription factor in governing proper pancreatic β-cell differentiation as well as maintaining a normal β-cell phenotype, including its growth and secretory functions, a deficiency of this factor will result in a significant reduction of β-cell mass and an impaired insulin-secretory response upon glucose challenge (Hart et al., 2000). Of particular interest in this context is the genetically modified *db/db* mouse, frequently used in T2DM research, which has had its *lep* genes knocked out. The resultant absence of leptin disrupts normal appetite regulation and energy intake. As adults, *db/db* mice become obese and develop obesity-induced T2DM. Such animal model is often associated with features manifested in typical patients with T2DM, including neuropathy and cardioregulatory dysfunctions, which enables it to be suitable for studying these diabetic complications (Goncalves et al., 2009).

The Zucker Diabetic Fatty (ZDF) rat has also been used as representative model for T2DM in rats. Genetic predisposition or general phenotype of ZDF rats than makes them a good model (Finegood et al., 2001). This diabetes-prone strain of animals, also with defects in their leptin receptors, has an obviously enhanced β-cell apoptosis that could possibly be attributed to lipotoxicity (Shimabukuro et al., 1998). Different with this obese ZDF rat, another common model, the Goto-Kakizaki (GK) rat, is a lean model of T2DM. These rats are characterized by a defective β-cell mass when young who progressively result in a markedly impaired glucose tolerance at the adult stage (Briaud et al., 2002).

A high-calorie diet is commonly administered in order to induce the diabetic condition; in some cases, low doses of STZ treatment are used to achieve similar results without rendering any significant damages to islet morphology (Okamoto et al., 2008). STZ can also be used to induce T2DM in neonatal rodents. In this method, a mild and stable form of T2DM is produced by a single dose of STZ (90 mg/kg) in 2-day-old neonatal mice or rats. This approach can induce a limited degree of β-cell damage followed by limited regeneration of islet cells as the animals grow, ultimately resulting in β-cell secretory dysfunction in the adult animals (Bonnevie-Nielsen et al., 1981). This can be a good model for studying what the regenerative mechanisms of the defective pancreas are and how the promotion of β-cell neogeneis is like by activating the undifferentiated progenitors. Diabetes models have also been developed in fish. The zebrafish, in particular, has been widely used in pancreatic development and regeneration research owing to its amenability to direct observation of its embryonic development under a microscope (Kinkel & Prince, 2009). Additionally, research in the dogfish has yielded some novel dogfish-derived peptides reported to have anti-diabetic effects (Huang & Wu, 2005).

7.2.3 Animal Models of Pancreatitis

There are both surgical and non-surgical methods by which to produce experimental models of acute pancreatitis (AP). A critical appraisal of recent advances in basic research in AP can be found in our recent review article (Chan & Leung, 2007). Figure 7.2 summarizes the most commonly used animal models of AP. Non-surgical, or drug induction, methods include, but are not limited to, caerulein-arginine-, and choline-deficient ethionine diet-induced approaches. Among them, a peptide analogue of the digestive enzyme cholecystokinin (CCK), called caerulein has been frequently used to induce AP in experimental animals. Two intra-peritoneal injections of a supraphysiological dose of caerulein are able to achieve a mild form of AP while multiple doses of caerulein can induce a severe form (Chan & Leung, 2006). The success rate of caerulein-induced AP in terms of degree of pancreatic injury and damage is normally assessed in terms of elevation of plasma α-amylase and lipase levels, as well as through evaluation of pancreatic edema and pancreatic histopathology, hours after administration of caerulein.

Surgical induction methods include, to name but a few, vascular-induced AP, duct obstruction-induced AP, closed duodenal loop-induced AP, and duct infusion-induced AP. The duct obstruction-induced model is especially common in AP research, requiring the relatively simple procedure of double ligation of the common biliopancreatic duct near the duodenal wall (Chan & Leung, 2007). This model

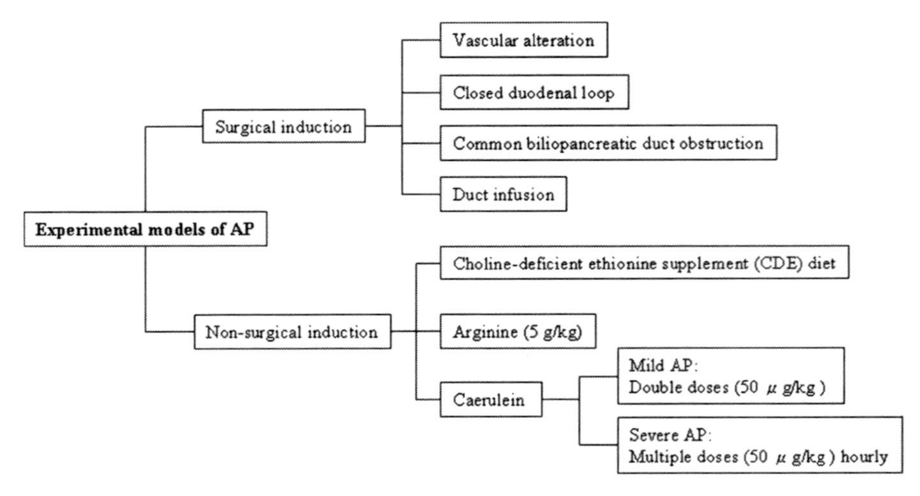

Fig. 7.2 A summary of the common experimental models of acute pancreatitis (AP) in basic pancreatic research

focuses on the etiology of AP as it mimics gallstone obstruction-induced pancreatitis in the clinical setting. It is interesting to note that the pancreatitis condition in this model could be completely reversed by removal of the blockade after a 24-h occlusion period (Azima et al., 1996). Simply put, its advantages have made this model a favourite for the studies of AP physiology and pathophysiology as well as studies of therapeutic treatments for the clinically relevant condition of gallstone-induced AP.

7.2.4 Animal Models of Pancreatic Cancer

Nude mice have been an important xenograft model for studying human pancreatic cancers in terms of their development, metastasis or response to different pharamacological agents (Shi & Xie, 2000). Both the subcutatneous injection of a pancreatic cancer cell suspension and, more commonly, the surgical orthotopic implantation of human pancreatic tumour fragments into the mouse pancreas, have been applied in many pancreatic cancer studies (Kim et al., 2008; Chen et al., 2009; Pérez-Torras et al., 2008). The latter model allows spontaneous widespread metastasis of the cancer cells to occur within a short time period, though the metastasis potential is often cell line dependent (Garofalo et al., 1993). Incorporation of green fluorescent protein into cancer cells or tumours can facilitate real-time visual imaging and assessment of tumour growth in vivo.

There are also many laboratory carcinogens available to chemically induce pancreatic cancers in animals. The organ specific carcinogen, 7,12-dimethylbenz-(a)anthracene (DMBA), has been used extensively to induce pancreatic adenocarcinomas by implanting it directly into the pancreas of Sprague-Dawley rats (Longnecker, 1990). While DMBA-induction method may require a longer period for tumourigenesis to occur, is the resultant tumours are relatively accessible for growth examination during drug treatment compared with tumours generated in surgical induction models.

7.3 Islet and Acinar Cell Isolation

A substantial advancement in the pancreatic research is owed to the development of islet and acinar cell isolation methods. The isolation of islets, in particular, greatly facilitates ex vivo functional studies, allowing more direct in vivo translation compared other β-cell line studies. Recently, substantial efforts have been put into harvesting islets with enhanced improvement in purity, viability and functionality. Optimization of the conditions for enzymatic digestion of pancreatic tissues often dictates the success of the isolation process. The typical procedures for islet isolation are schematically illustrated in Fig. 7.3. There have been numerous reports comparing the isolation outcomes achieved using different collagenases with various proteases (Brandhorst et al., 2003, 2005). A typical digestion protocol calls for intra-ductal administration of collagenase so as to achieve a more thorough infusion and digestion process; however, the conditions for such digestion need

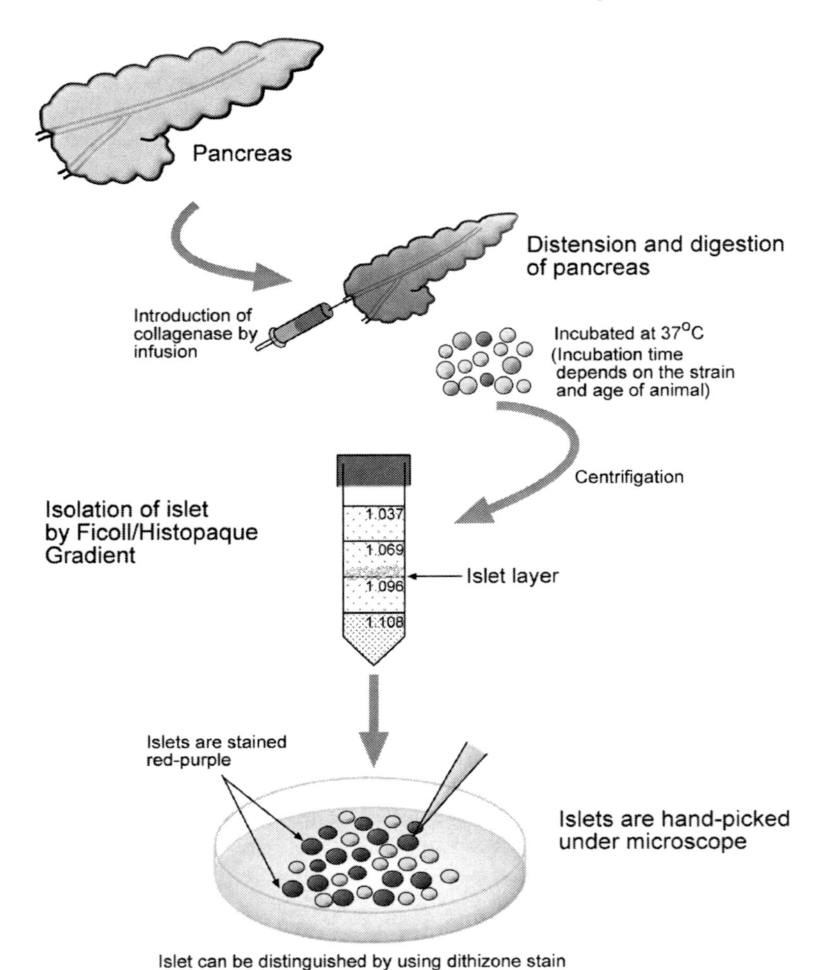

Fig. 7.3 A schematic diagram illustrating the basic experimental procedures for islet isolation

further optimization to prevent intra-islet penetration of the enzymes (Cross et al., 2008). Supplementary reagents (i.e., Ficoll or Histopaque) have been included to separate cell types, greatly facilitating islet isolation in digested tissues by density gradient (Anderson et al., 2007; Ohgawara et al., 1998). Despite this, complete removal of the peri-acinar fibrous tissue remains the limiting factor for harvesting of islets.

Various islet purification strategies have been applied. For example, dithiozone, an established dye for chelating zinc in islets, can produce a red-purplish colour to assist in the identification of the islets (McNary, 1954). Some new digestion chambers have also been developed to assist in mechanical dissociation or filtering out of unwanted components (Gray et al., 2004; Dufrane et al., 2005). In addition to harvesting islets, recent studies have reported protocols to purify β-cells from islets, based on either β-cell specific molecules (Banerjee & Otonkoski, 2009) or manipulation of their cellular metabolisms followed by flow cytometry (Smelt et al., 2008).

The protocols for the pancreatic acinar cell isolation have been well described previously (Schulz et al., 1988). The process appears to be indifferent to which typical enzyme digestion procedure is used. One major obstacle is maintaining the survival of the isolated acinar cells. Decreases in amylase and lipase activity over time in culture are also commonly noted. Optimization of the culture conditions and nutritional set-up is important for maintenance of acinar cell viability for long-term experiments (Kurup & Bhonde, 2002; Singh et al., 2008). Interestingly, several cytokines become upregulated during the isolation of acinar cells from pancreatic tissues via NF-κB and p38 MAP kinase. This observation appears to be due, at least in part, to their blunt response to some other physiological stimulus in culture (Blinman et al., 2000).

7.4 Islet Transplantation

The islet transplantation procedure has been a critical breakthrough in treating patients with T1DM and some severe forms of T2DM with chronic complications. The recent success of the Edmonton protocol for islet transplantation has brought about a new era for adequate control of euglycemic status, though insulin independence may not be sustained permanently (Shapiro et al., 2006). Current limitations include the fact that the transplanted islet grafts gradually lose their secretory function and thus may ultimately succumb to failure due to inadequate revascularization or allograft host immune rejection (Rickels et al., 2007). The inner core of the islets, constituted predominantly by insulin-secreting β-cells, is especially susceptible to insufficiency of oxygen and nutrients (Vasir et al., 1998). There have been various strategies developed to maintain β-cell graft function and islet survival in vivo, leading to improvement in islet transplantation (Narang & Mahato, 2006). Different experimental animal models of islet transplantation have been developed as a platform for the assessment of the outcome from various manipulations and monitoring host immune response.

7.4.1 Animal Models of Islet Transplantation

Transplantation studies, especially of xenotransplantation (between species), require the use of immunodeficient animals in order to avoid acute graft rejections. The most common host animal model is the nude mouse, a genetic mutant that thus has an inhibited immune system due to its lack of a thymus. Most nude mice, such as *BALB/c* nude and *NU/NU* mice, are T-cell deficient, having developed with only a B-cell system. On the other hand, the Severe Combined Immunodeficient (SCID) mice have a total immature cellular or humoral immune system. These mice fail to develop any B-cells or T-cells to activate the compliment system. SCID mice are thus widely used as a tool to host transplanted tissues in different studies. The SCID mutation has also been developed on the NOD background, creating the NOD-SCID mouse used in diabetes research. Many researchers choose to perform human islet transplantation studies into the diabetic NOD-SCID mouse model. Some laboratories have even humanized these mice by engraftment of human blood cells (Brands et al., 2008).

Unlike other tissue or organ transplantations, islets are not transplanted homotropically into the highly sensitive pancreas, but instead are transplanted into other heterotropic graft sites. Clinical islet transplantation is normally performed through infusion of the isolated islets via the hepatic portal vein into the liver. Yet studies using transplanted mouse models have revealed profound functional impairment in islet grafts retrieved from the liver. These retrieved islets were found to have markedly depleted expression of several β-cell phenotypic factors as well as low insulin release in response to a glucose challenge. These observations have drawn the attention to the notion of implantation site-dependent functional performance of the transplanted islet grafts (Mattsson et al., 2004; Lau et al., 2007). In view of the fact that the loss of islet graft functions might be partly attributable to the necessity for multiple islet donors to cure each individual T1DM patient, researchers are working to find an optimal anatomical site for islet transplantation (van der Windt et al., 2008). New islet transplantation sites, such as the renal capsule, spleen, or the omental pouch, have been reported in animal studies (Merani et al., 2008). The route of each transplantation method offers particular benefits. In the clinical setting, subcutaneous transplantation maximizes patient safety. In our laboratory, we have established the renal capsule as an experimental transplantation site; a brief flow of procedures used in this approach is illustrated in Fig. 7.4.

While transplantation studies usually involve intra-species tissue grafting, recent advances have also suggested utilization of islets from xenogenic donors (Schuurman & Pierson, 2008; Cozzi & Bosio, 2008). This idea has been an important breakthrough for dealing with the limited availability of human donor islets. Porcine islets have received substantial attention, particularly since it was demonstrated that they respond to glucose challenge in the same physiological ranges as do human islets (Dufrane & Gianello, 2008). Furthermore, porcine islets may perform better than islets from other species, such as the rat, in terms of achieving hyperglycemia reversal (Davalli et al., 1995).

7.4.2 Recent Advances in Islet Transplantation

Minimization of acute rejection of grafted tissues is probably the most critical issue in achieving successful clinical transplantation. This issue is being addressed by studies of immuno-isolation of transplanted islets by microencapsulation. Encapsulation of islets with biomaterials can provide a physical barrier while allowing the passage of glucose and insulin and, at the same time, preventing the

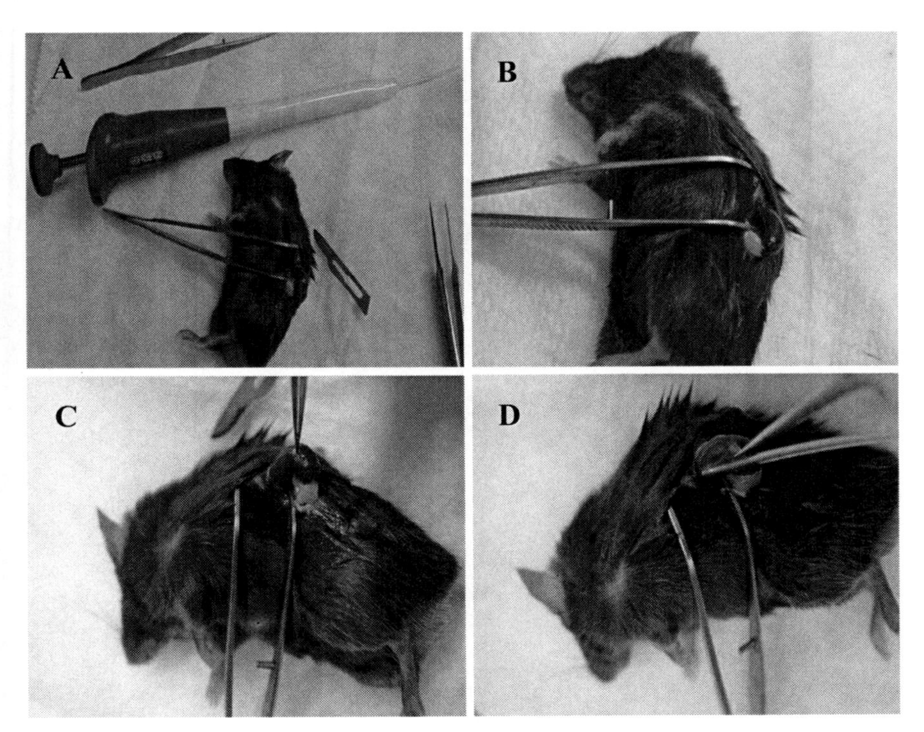

Fig. 7.4 The photographs showing the procedures of islet transplantation in mouse. **a** The mouse was first anesthetized. **b** The left abdomen was cut open to expose the left kidney. **c** A small incision was made on the renal capsule. **d** Islets were injected between the renal parenchyma and capsule using a positive displacement pipette (withdrawn from Suen, 2007)

entry of large molecules like antibodies (Campos-Lisbôa et al., 2008; Teramura et al., 2007). This strategy is designed to greatly maintain islet graft survival in the host, thus facilitating xenogenic-based therapeutic options. A common protocol with slight variations for islet encapsulation has been reported in many studies (Foster et al., 2007; Calafiore et al., 2006) and the procedures involved are illustrated in Fig. 7.5. Typical protocols make use of the alginate-based materials for microencapsulation, titrated against a cation solution (usually calcium or barium) for gelification and beads formation. Biocompatibility of the encapsulation materials is the major concern for success, and in this context, alginate is a suitable choice for not interfering with the celluar functions of the transplanted islets (Fritschy et al., 1991). Recent findings have even suggested that immunosupression might not be necessary to maintain long-term normoglycemia in diabetic animals given xenogenic islet transplantations (Meyer et al., 2008). These data provide preliminary support to the possibility that immunosuppressive drugs may be minimized after islet transplantation in patients with diabetes.

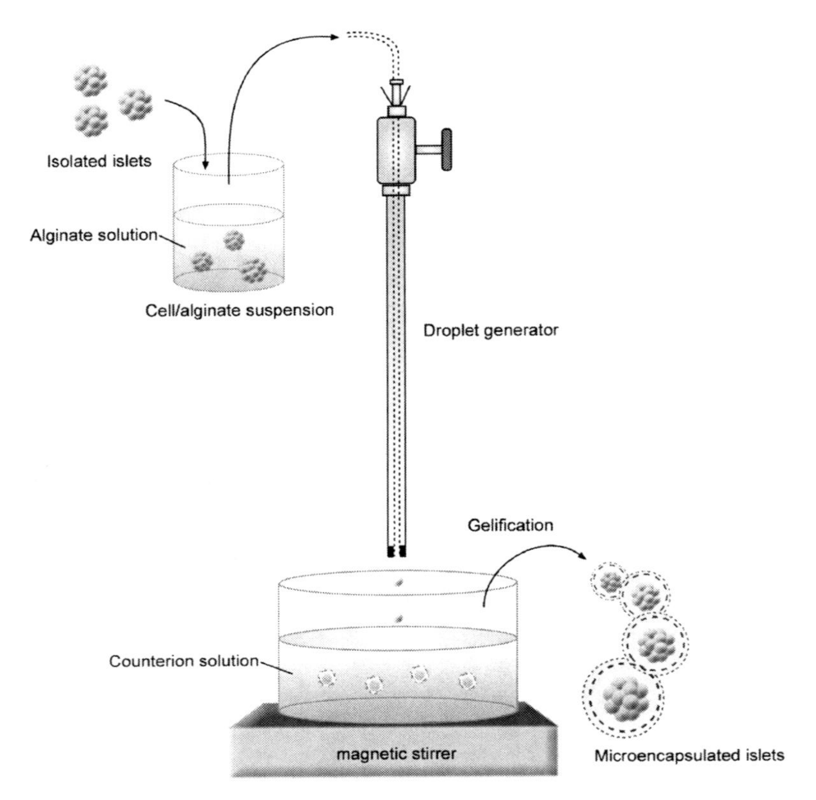

Fig. 7.5 A schematic diagram depicting the standard procedures for islet microencapsulation

Various pre-transplantation biological manipulations of the islets have also been suggested to optimize protocols for feasible islet grafting. Gene delivery of the vascular endothelial growth factor (VEGF), for example, in isolated human islets can boost revascularization in transplanted diabetic SCID mice models (Narang et al., 2004); results from glucose tolerance tests also revealed better performance in ameliorating hyperglycemia in animals transplanted with VEGF-transfected islets (Chae et al., 2005). Chemical-induced transfection would generally be preferable to viral vector in gene delivery in clinical applications, as the latter may affect cellular immunogenicity or induce chemokine production.

7.5 Expression and Functional Studies

Commonly employed experimental techniques for pancreatic research include, but are not limited to, molecular and functional studies which are summarized in Fig. 7.6.

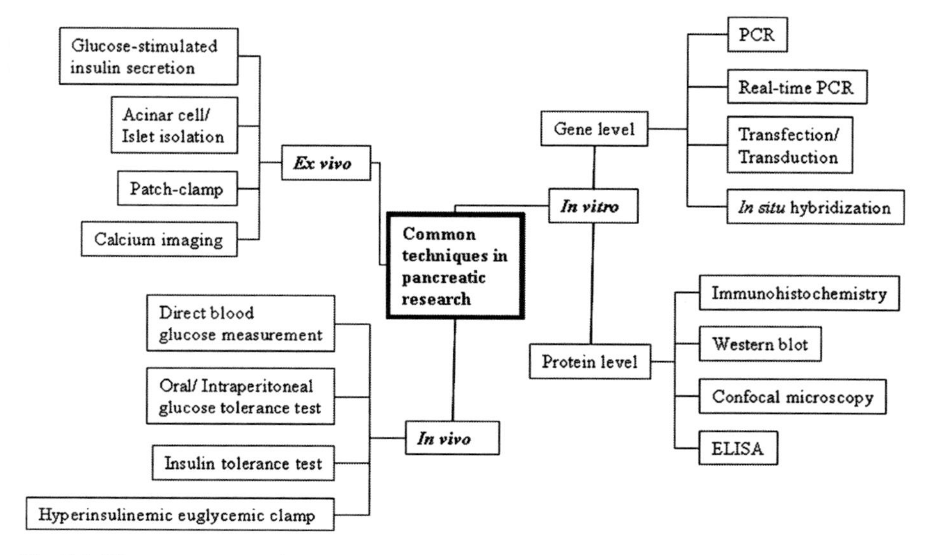

Fig. 7.6 The common experimental techniques recruited in basic pancreatic research

7.5.1 Gene expression studies

In terms of molecular techniques, expression studies of a particular gene at the mRNA level are indispensable. Polymerase chain reaction (PCR) is a typical strategy used to compare gene expression levels in a semi-quantitative manner. Quantitative real-time PCR can be applied to measure accumulation of gene products of interest by analysis of a fluorophore during the exponential stages of the PCR. The threshold cycle (C_T), when the amplified target genes are first detected, are compared and the exact fold change between different treatment groups can thus be obtained. Two common dyes used in real-time PCR are Sybr Green, which tags double-stranded DNAs, and the Taqman probe, which releases fluorophores only when a specific gene sequence is amplified. Precise localization of specific genetic transcripts in cells or tissues requires use of in situ hybridization methodology, which directly labels gene transcripts of interest either in tissues sections or in whole mounts by means of complementary nucleotide binding.

7.5.2 Protein Expression Studies

Studies of protein expression levels are also indispensable for pancreatic research. Western blotting (Burnette, 1981) employing antibodies for particular proteins of interest is the most common method used to assess protein expression levels. Protocols for the isolation of membrane or nuclear proteins have been well developed for the analysis of numerous proteins. If quantification of a protein or peptide is required, the enzyme-linked immunosorbent assay (ELISA) methodology (Lequin, 2005) should be used in which the amount of the target protein or peptide in the sample is proportionally reflected by a colorimetric method. The concept of antigen-antibody binding has been further elaborated and applied in a technique named flow cytometry. This fluorescence-based technique is useful in terms of allowing simultaneous detection and quantification of multiple antigens of interest. The data generated, in addition, allow a gated strategy in which a particular portion of cells within a population is analysed. A specialized type of flow cytometry, named the fluorescence-activated cell sorting (FACS), is a powerful technique which allows the purification of a particular cell type of interest within a pool of cell population based on its specialized antigen expression. This has been widely applied in different stem cell research (Alexander et al., 2009). Localization of target proteins within a cell or tissue can be achieved by immunocytochemistry (ICC) or immunohistochemistry (IHC), respectively. Specimens, either paraffin-embedded or cryo-preserved, are fixed and labelled through incubation with specific antibodies against the proteins of interest. Immunoreactivity is reflected in staining or fluorescence intensity, thus allowing an assessment of protein expression levels as well. Co-application of two or more antibodies can be used to demonstrate simultaneous expression of proteins within cells or tissues. Multiple-protein ICC and IHC is best achieved with fluorescent labelling protocols, which requires a fluorescent microscope for detection and imaging. A confocal imaging system further allows three-dimensional reconstruction of tissues, such as whole-mount pancreas tissue. Two-photon microscopy, a viable alternative to traditional confocal microscopy due to its lower phototoxicity and deeper tissue penetration, was recently demonstrated to depict the dynamics of the exocytotic events in a β-cell within a mouse pancreatic islet (Takahashi et al., 2002).

7.5.3 Transfection

Genetic modification requires transfection and transduction techniques. These approaches enable foreign nucleic acids to be introduced into host cells through various non-viral and viral vehicles, respectively. Different with employing viral particles as vectors, typical non-viral transfection is a chemical-based method that allows certain nucleic acid sequences to directly enter the cell cytoplasm through alteration of the cell membrane. A commonly used reagent is the Lipofectamine. Most transfected/transduced genes may only be transiently expressed, but some viral vectors (e.g. lentiviruses) allow more stable transduction due to their ability to incorporate genes into the host genome, thus being replicated in host cell mitosis. If a stable and long-term transduction is aimed, the genes are often transduced with a resistance gene to particular antibiotics. Selection of the successfully transduced cells by those antibiotics after the transduction procedures allows a thorough screening and the propagation of the transduced-only cells. The RNA interference (RNAi) technique, especially using short interfering RNAs (siRNAs), is commonly used in gene knockdown experiments. RNAi impedes the expression of a specific gene by annealing to their mRNA sequences and thereby marking them for enzymatic degradation. Recent studies also make use of the small-haripin RNA/short hairpin RNA (shRNA) for achieving a gene knockdown. The shRNA sequences contain a tight hairpin turn that allows an increased stability of the structure and

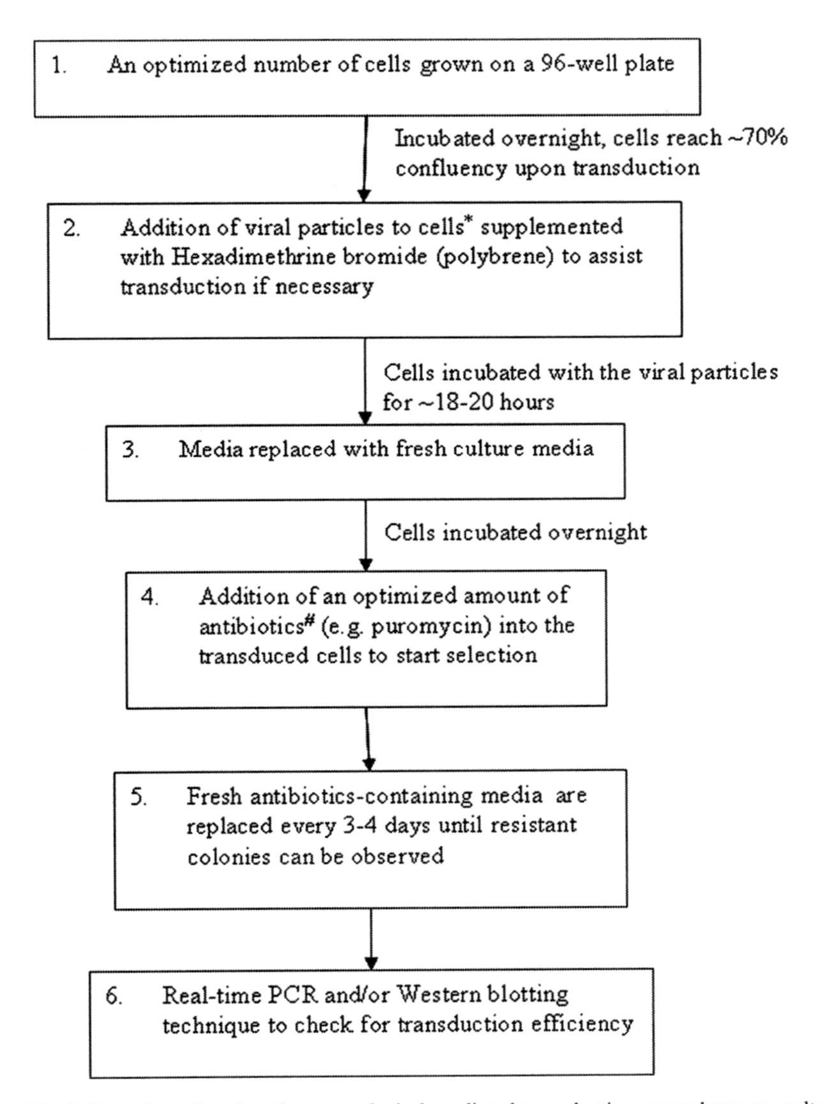

Fig. 7.7 A flow chart showing the general viral-mediated transduction procedures on cultured cells. *A range of multiplicity of infection (MOI; usually 0.5, 1, 2, and 5) should be tested to determine the optimal transduction efficiency. #A killing curve of the antibiotics for each specific cell type should be tested beforehand to determine the minimal amount of antibiotics used to kill all untransduced cells

achieves a more stable knockdown efficiency. These processes can be done both in vitro in cell culture system and in vivo within targeted organ systems, as recently described in the literature (Larson et al., 2007). A general viral-mediated transduction procedure on cultured cells is illustrated in Fig. 7.7. Verification of a successful transduction experiment is normally done by transducing the green fluorescence protein (GFP)-tagged viruses into cells. Representative images of the human fetal pancreas-derived progenitor cells from our laboratory transduced with the GFP viral particles are shown as an example in Fig. 7.8.

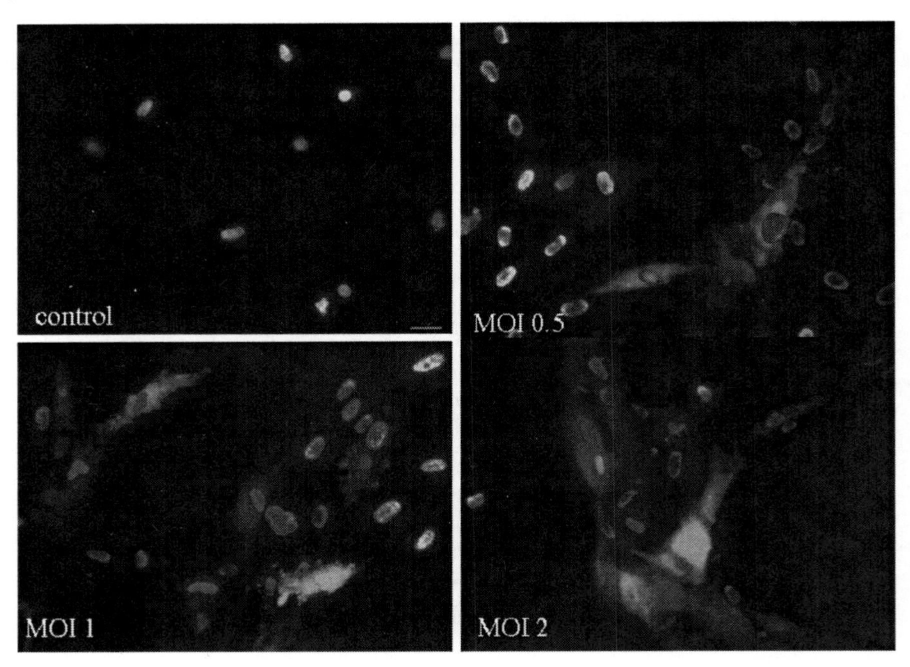

Fig. 7.8 The representative images showing the human fetal pancreas-derived progenitor cells transduced with the *green* fluorescence protein-tagged lentiviral particles in different multiplicity of infection (MOI). Note that the transduction efficiency (percentage of transduced cells) is different under different MOI. *Blue*: DAPI; *green*: viral GFP. Magnification ×630, scale bar =40 μm (For interpretation of the references to colour in this figure legend, please be referred to the online version.)

7.5.4 Functional Studies

In vitro (or ex vivo) functional studies, which have frequently been applied in pancreatic research, entail morphological and functional assessments of an isolated islet. This approach allows detailed examination of pancreatic islet function and β-cell structure that may be affected by different risk factors such as glucotoxicity, lipotoxicity, hypoxia or oxidative stress. The physiological response of an islet to glucose-stimulated insulin secretion is often studied as a functional correlate of in vivo physiology. Other insulin secretagogues (i.e., like amino acids, glyceraldehyde, and K^+ ions) are also commonly used to test islet secretory capacity. ELISA is used to quantify secretion of peptides of interest, such as (pro)insulin or C-peptide, as well as to assess the extent of cell proliferation or cell death, using bromodeoxyuridine and cell death ELISAs, respectively. Since normal glucose-stimulated insulin secretion (Define GSIS) depends on normal functioning of islet ion channels, the patch-clamp technique can also be applied to examine the channel activities, thereby reflecting ion movement dynamics (Marigo et al., 2009).

Several typical parameters should be examined and recorded in in vivo functional studies of the pancreas; they include animal survival, body weight, and water consumption. Blood samples are usually collected to determine glucose or insulin levels. The oral glucose tolerance test (OGTT) or intraperitoneal glucose tolerance test (IGTT) are also common methods used to assess the animals' abilities to reverse hyperglycemia. In these assays, glucose solution is either gavaged or injected, and blood glucose levels are then monitored within 240 min. The animals' insulin sensitivity can be assessed by an insulin tolerance test wherein insulin is administered intraperitoneally and subsequent changes in blood glucose levels are monitored. Insulin resistance can be measured and quantified using a hyperinsulinemic euglycemic clamp (Wilkes et al., 2009). In this method, insulin is continuously perfused into the animal and the amount of glucose required to compensate for the increased insulin levels without causing hypoglycaemia is carefully examined and recorded. Greater glucose supplementation requirement indicates greater insulin-sensitivity.

References

Alexander CM, Puchalski J, Klos KS, Badders N, Ailles L, Kim CF, Dirks P and Smalley MJ. Separating stem cells by flow cytometry: reducing variability for solid tissues. *Cell Stem Cell* **5**:579–583, 2009.

Anderson JM, Deeds MC, Armstrong AS, Gastineau DA and Kudva YC. Utilization of a test gradient enhances islet recovery from deceased donor pancreases. *Cytotherapy* **9**:630–636, 2007.

Arora S, Ojha SK and Vohora D. Characterisation of Streptozotocin Induced Diabetes Mellitus in Swiss Albino Mice. *Global Journal of Pharmacology* **3**:81–84, 2009.

Atkinson MA and Leiter EH. The NOD mouse model of type 1 diabetes: as good as it gets? *Nat Med* **5**:601–604, 1999

Azima B, Kao RL, Youngberg G, Williams D and Browder W. A new animal model of reversible acute pancreatitis. *J Surg Res* **63**:419–424, 1996.

Banerjee M and Otonkoski T. A simple two-step protocol for the purification of human pancreatic beta cells. *Diabetologia* **52**:621–625, 2009.

Blinman TA, Gukovsky I, Mouria M, Zaninovic V, Livingston E, Pandol SJ and Gukovskaya AS. Activation of pancreatic acinar cells on isolation from tissue: cytokine upregulation via p38 MAP kinase. *Am J Physiol Cell Physiol* **279**:C1993–C2003, 2000.

Bonnevie-Nielsen V, Steffes MW and Lernmark A. A major loss in islet mass and B-cell function precedes hyperglycemia in mice given multiple low doses of streptozotocin. *Diabetes* **30**: 424–429, 1981.

Brandhorst H, Brandhorst D, Hesse F, Ambrosius D, Brendel M, Kawakami Y and Bretzel RG. Successful human islet isolation utilizing recombinant collagenase. *Diabetes* **52**:1143–1146, 2003.

Brandhorst H, Brendel MD, Eckhard M, Bretzel RG and Brandhorst D. Influence of neutral protease activity on human islet isolation outcome. *Transplant Proc* **37**:241–242, 2005.

Brands K, Colvin E, Williams LJ, Wang R, Lock RB and Tuch BE. Reduced immunogenicity of first-trimester human fetal pancreas. *Diabetes* **57**:627–634, 2008.

Briaud I, Kelpe CL, Johnson LM, Tran PO and Poitout V. Differential effects of hyperlipidemia on insulin secretion in islets of langerhans from hyperglycemic versus normoglycemic rats. *Diabetes* **51**:662–668, 2002.

Brissova M, Shiota M, Nicholson WE, Gannon M, Knobel SM, Piston DW, Wright CV and Powers AC. Reduction in pancreatic transcription factor PDX-1 impairs glucose-stimulated insulin secretion. *J Biol Chem* **277**:11225–11232, 2002.

Burnette WN. "Western blotting": electrophoretic transfer of proteins from sodium dodecyl sulfate–polyacrylamide gels to unmodified nitrocellulose and radiographic detection with antibody and radioiodinated protein A. *Anal Biochem* **112**:195–203, 1981.

Calafiore R, Basta G, Luca G, Lemmi A, Racanicchi L, Mancuso F, Montanucci MP and Brunetti P. Standard technical procedures for microencapsulation of human islets for graft into non-immunosuppressed patients with type 1 diabetes mellitus. *Transplant Proc* **38**:1156–1157, 2006.

Campos-Lisbôa AC, Mares-Guia TR, Grazioli G, Goldberg AC and Sogayar MC. Biodritin microencapsulated human islets of Langerhans and their potential for type 1 diabetes mellitus therapy. *Transplant Proc* **40**:433–435, 2008.

Castaing M, Duvillié B, Quemeneur E, Basmaciogullari A and Scharfmann R. Ex vivo analysis of acinar and endocrine cell development in the human embryonic pancreas. *Dev Dyn* **234**: 339–345, 2005.

Cefalu WT. Animal models of type 2 diabetes: clinical presentation and pathophysiological relevance to the human condition. *ILAR J* **47**:186–198, 2006.

Chae HY, Lee BW, Oh SH, Ahn YR, Chung JH, Min YK, Lee MS, Lee MK and Kim KW. Effective glycemic control achieved by transplanting non-viral cationic liposome-mediated VEGF-transfected islets in streptozotocin-induced diabetic mice. *Exp Mol Med* **37**:513–523, 2005.

Chan YC and Leung PS. AT1 receptor antagonism ameliorates acute pancreatitis-associated pulmonary injury. *Regul Pept* **134**:46–53, 2006.

Chan YC and Leung PS. Acute pancreatitis: animal models and recent advances in basic research. *Pancreas* **34**:1–14, 2007.

Chan YC and Leung PS. Angiotensin II type 1 receptor-dependent nuclear factor-kappaB activation-mediated proinflammatory actions in a rat model of obstructive acute pancreatitis. *J Pharmacol Exp Ther* **323**:10–18, 2007

Chan YC and Leung PS. Involvement of redox-sensitive extracellular-regulated kinases in angiotensin II-induced interleukin-6 expression in pancreatic acinar cells. *J Pharmacol Exp Ther* **329**:450–458, 2009.

Chen H, Sun B, Pan S, Jiang H and Sun X. Dihydroartemisinin inhibits growth of pancreatic cancer cells in vitro and in vivo. *Anticancer Drugs* **20**:131–140, 2009.

Chen J, Chen Z, Chen M, Li D, Li Z, Xiong Y, Dong J and Li X. Role of fibrillar Tenascin-C in metastatic pancreatic cancer. *Int J Oncol* **34**:1029–1036, 2009.

Chuang JC, Cha JY, Garmey JC, Mirmira RG and Repa JJ. Research resource: nuclear hormone receptor expression in the endocrine pancreas. *Mol Endocrinol* **22**:2353–2363, 2008.

Cozzi E and Bosio E. Islet xenotransplantation: current status of preclinical studies in the pig-to-nonhuman primate model. *Curr Opin Organ Transplant* **13**:155–158, 2008.

Cross SE, Hughes SJ, Partridge CJ, Clark A, Gray DW and Johnson PR. Collagenase penetrates human pancreatic islets following standard intraductal administration. *Transplantation* **86**: 907–911, 2008.

Davalli AM, Ogawa Y, Scaglia L, Wu YJ, Hollister J, Bonner-Weir S and Weir GC. Function, mass, and replication of porcine and rat islets transplanted into diabetic nude mice. *Diabetes* **44**:104–111, 1995.

Dufrane D, Goebbels RM, Guiot Y, Squifflet JP, Henquin JC and Gianello P. A simple method using a polymethylpenten chamber for isolation of human pancreatic islets. *Pancreas* **30**:e51–e59, 2005.

Dufrane D and Gianello P. Pig islet xenotransplantation into non-human primate model. *Transplantation* **86**:753–760, 2008.

Farrow B, Berger DH and Rowley D. Tumor-derived pancreatic stellate cells promote pancreatic cancer cell invasion through release of thrombospondin-2. *J Surg Res* **156**:155–160, 2009.

Finegood DT, McArthur MD, Kojwang D, Thomas MJ, Topp BG, Leonard T and Buckingham RE. Beta-cell mass dynamics in Zucker diabetic fatty rats. Rosiglitazone prevents the rise in net cell death. *Diabetes* **50**:1021–1029, 2001.

Foster JL, Williams G, Williams LJ and Tuch BE. Differentiation of transplanted microencapsulated fetal pancreatic cells. *Transplantation* **83**:1440–1448, 2007.

Fritschy WM, Wolters GH and van Schilfgaarde R. Effect of alginate-polylysine-alginate microencapsulation on in vitro insulin release from rat pancreatic islets. *Diabetes* **40**:37–43, 1991.

Garofalo A, Chirivi RG, Scanziani E, Mayo JG, Vecchi A and Giavazzi R. Comparative study on the metastatic behavior of human tumors in nude, beige/nude/xid and severe combined immunodeficient mice. *Invasion Metastasis* **13**:82–91, 1993.

Goncalves AC, Tank J, Diedrich A, Hilzendeger A, Plehm R, Bader M, Luft FC, Jordan J and Gross V. Diabetic hypertensive leptin receptor-deficient db/db mice develop cardioregulatory autonomic dysfunction. *Hypertension* **53**:387–392, 2009.

Gray DW, Sudhakaran N, Titus TT, McShane P and Johnson P. Development of a novel digestion chamber for human and porcine islet isolation. *Transplant Proc* **36**:1135–1138, 2004.

Guillemain G, Filhoulaud G, Da Silva-Xavier G, Rutter GA and Scharfmann R. Glucose is necessary for embryonic pancreatic endocrine cell differentiation. *J Biol Chem* **282**:15228–15237, 2007.

Hart AW, Baeza N, Apelqvist A and Edlund H. Attenuation of FGF signalling in mouse beta-cells leads to diabetes. *Nature* **408**:864–868, 2000.

von Herrath M and Nepom GT. Animal models of human type 1 diabetes. *Nat Immunol* **10**: 129–132, 2009.

Huang F and Wu W. Antidiabetic effect of a new peptide from Squalus mitsukurii liver (S-8300) in alloxan-diabetes. *Clin Exp Pharmacol Physiol* **32**:521–525, 2005.

Junod A, Lambert AE, Stauffacher W and Renold AE. Diabetogenic action of streptozotocin: relationship of dose to metabolic response. *J Clin Invest* **48**:2129–2139, 1969.

Kadowaki T. Insights into insulin resistance and type 2 diabetes from knockout mouse models. *J Clin Invest* **106**:459–465, 2000.

Kaneko Y, Kimura T, Taniguchi S, Souma M, Kojima Y, Kimura Y, Kimura H and Niki I. Glucose-induced production of hydrogen sulfide may protect the pancreatic beta-cells from apoptotic cell death by high glucose. *FEBS Lett* **583**:377–382, 2009.

Katz EB, Stenbit AE, Hatton K, DePinho R and Charron MJ. Cardiac and adipose tissue abnormalities but not diabetes in mice deficient in GLUT4. *Nature* **377**:151–155, 1995.

Kim H, Morgan DE, Buchsbaum DJ, Zeng H, Grizzle WE, Warram JM, Stockard CR, McNally LR, Long JW, Sellers JC, Forero A and Zinn KR. Early therapy evaluation of combined anti-death receptor 5 antibody and gemcitabine in orthotopic pancreatic tumor xenografts by diffusion-weighted magnetic resonance imaging. *Cancer Res* **68**:8369–8376, 2008.

Kinkel MD and Prince VE. On the diabetic menu: Zebrafish as a model for pancreas development and function. *Bioessays* **31**:139–152, 2009.

Kroon E, Martinson LA, Kadoya K, Bang AG, Kelly OG, Eliazer S, Young H, Richardson M, Smart NG, Cunningham J, Agulnick AD, D'Amour KA, Carpenter MK and Baetge EE. Pancreatic endoderm derived from human embryonic stem cells generates glucose-responsive insulin-secreting cells in vivo. *Nat Biotechnol* **26**:443–452, 2008.

Kurup S and Bhonde RR. Analysis and optimization of nutritional set-up for murine pancreatic acinar cells. *JOP* **3**:8–15, 2002.

Larson SD, Jackson LN, Chen LA, Rychahou PG and Evers BM. Effectiveness of siRNA uptake in target tissues by various delivery methods. *Surgery* **142**:262–269, 2007.

Lau J, Mattsson G, Carlsson C, Nyqvist D, Köhler M, Berggren PO, Jansson L and Carlsson PO. Implantation site-dependent dysfunction of transplanted pancreatic islets. *Diabetes* **56**: 1544–1550, 2007.

Lau ST, Lin ZX, Zhao M and Leung PS. Brucea javanica fruit induces cytotoxicity and apoptosis in pancreatic adenocarcinoma cell lines. *Phytother Res* **22**:477–486, 2008.

Leiter EH, Malinoski FJ and Eppig JJ. An epithelial cell line with chronic polyoma infection established from a spontaneous mouse pancreatic adenocarcinoma. *Cancer Res* **38**:969–977, 1978.

Leiter EH, Prochazka M and Coleman DL. The non-obese diabetic (NOD) mouse. *Am J Pathol* **128**:380–383, 1987.

Lenzen S and Panten U. Alloxan: history and mechanism of action. *Diabetologia* **31**:337–342, 1988.

LeRoith D and Gavrilova O. Mouse models created to study the pathophysiology of Type 2 diabetes. *Int J Biochem Cell Biol* **38**:904–912, 2006

Lequin RM. Enzyme immunoassay (EIA)/enzyme-linked immunosorbent assay (ELISA). *Clin Chem* **51**:2415–2418, 2005.

Leung KK and Leung PS. Effects of hyperglycemia on angiotensin II receptor type 1 expression and insulin secretion in an INS-1E pancreatic beta-cell line. *JOP* **9**:290–299, 2008.

Longnecker D. Experimental pancreatic cancer: role of species, sex and diet. *Bull Cancer* **77**: 27–37, 1990.

Marigo V, Courville K, Hsu WH, Feng JM and Cheng H. TRPM4 impacts on Ca(2+) signals during agonist-induced insulin secretion in pancreatic beta-cells. *Mol Cell Endocrinol* **299**:194–203, 2009.

Masiello P. Animal models of type 2 diabetes with reduced pancreatic beta-cell mass. *Int J Biochem Cell Biol* **38**:873–893, 2006.

Mattsson G, Jansson L, Nordin A, Andersson A and Carlsson PO. Evidence of functional impairment of syngeneically transplanted mouse pancreatic islets retrieved from the liver. *Diabetes* **53**:948–954, 2004.

McNARY WF Jr. Zinc-dithizone reaction of pancreatic islets. *J Histochem Cytochem* **2**:185–194, 1954.

Merani S, Toso C, Emamaullee J and Shapiro AM. Optimal implantation site for pancreatic islet transplantation. *Br J Surg* **95**:1449–1461, 2008.

Meyer T, Höcht B and Ulrichs K. Xenogeneic islet transplantation of microencapsulated porcine islets for therapy of type I diabetes: long-term normoglycemia in STZ-diabetic rats without immunosuppression. *Pediatr Surg Int* **24**:1375–1378, 2008.

Murray HE, Paget MB, Bailey CJ and Downing R. Sustained insulin secretory response in human islets co-cultured with pancreatic duct-derived epithelial cells within a rotational cell culture system. *Diabetologia* **52**:477–485, 2009.

Mwangi S, Anitha M, Mallikarjun C, Ding X, Hara M, Parsadanian A, Larsen CP, Thule P, Sitaraman SV, Anania F and Srinivasan S. Glial cell line-derived neurotrophic factor increases beta-cell mass and improves glucose tolerance. *Gastroenterology* **134**:727–737, 2008.

Narang AS, Cheng K, Henry J, Zhang C, Sabek O, Fraga D, Kotb M, Gaber AO and Mahato RI. Vascular endothelial growth factor gene delivery for revascularization in transplanted human islets. *Pharm Res* **21**:15–25, 2004.

Narang AS and Mahato RI. Biological and biomaterial approaches for improved islet transplantation. *Pharmacol Rev* **58**:194–243, 2006.

Ohgawara H, Shikano T, Fukunaga K, Yamagishi M and Miyazaki S. Establishment of monolayer culture of pig pancreatic endocrine cells by use of nicotinamide. *Diabetes Res Clin Pract* **42**: 1–8, 1998.

Okamoto T, Kanemoto N, Ohbuchi Y, Okano M, Fukui H and Sudo T. Characterization of STZ-induced type 2 diabetes in Zucker fatty rats. *Exp Anim* **57**:335–345, 2008.

Onoue S, Hanato J and Yamada S. Pituitary adenylate cyclase-activating polypeptide attenuates streptozotocin-induced apoptotic death of RIN-m5F cells through regulation of Bcl-2 family protein mRNA expression. *FEBS J* **275**:5542–5551, 2008.

Ou D, Wang X, Metzger DL, Robbins M, Huang J, Jobin C, Chantler JK, James RF, Pozzilli P, Tingle AJ. Regulation of TNF-related apoptosis-inducing ligand-mediated death-signal pathway in human beta cells by Fas-associated death domain and nuclear factor kappaB. *Hum Immunol* **66**:799–809, 2005.

Parsa I and Marsh WH. Long-term organ culture of embryonic rat pancreas in a chemically defined medium. *Am J Pathol* **82**:119–128, 1976.

Pérez-Torras S, García-Manteiga J, Mercadé E, Casado FJ, Carbó N, Pastor-Anglada M and Mazo A. Adenoviral-mediated overexpression of human equilibrative nucleoside transporter 1 (hENT1) enhances gemcitabine response in human pancreatic cancer. *Biochem Pharmacol* **76**:322–329, 2008.

Petersen HV, Jørgensen MC, Andersen FG, Jensen J, F-Nielsen T, Jørgensen R, Madsen OD and Serup P. Pax4 represses pancreatic glucagon gene expression. *Mol Cell Biol Res Commun* **3**:249–254, 2000.

Rickels MR, Kamoun M, Kearns J, Markmann JF and Naji A. Evidence for allograft rejection in an islet transplant recipient and effect on beta-cell secretory capacity. *J Clin Endocrinol Metab* **92**:2410–2414, 2007.

Schnedl WJ, Ferber S, Johnson JH and Newgard CB. STZ transport and cytotoxicity. Specific enhancement in GLUT2-expressing cells. *Diabetes* **43**:1326–1333, 1994.

Schulz HU, Letko G, Spormann H, Sokolowski A and Kemnitz P. An optimized procedure for isolation of rat pancreatic acinar cells. *Anat Anz* **167**:141–150, 1988.

Schuurman HJ and Pierson RN 3rd. Progress towards clinical xenotransplantation. *Front Biosci* **13**:204–220, 2008.

Shapiro AM, Ricordi C, Hering BJ, Auchincloss H, Lindblad R, Robertson RP, Secchi A, Brendel MD, Berney T, Brennan DC, Cagliero E, Alejandro R, Ryan EA, DiMercurio B, Morel P, Polonsky KS, Reems JA, Bretzel RG, Bertuzzi F, Froud T, Kandaswamy R, Sutherland DE, Eisenbarth G, Segal M, Preiksaitis J, Korbutt GS, Barton FB, Viviano L, Seyfert-Margolis V, Bluestone J and Lakey JR. International trial of the Edmonton protocol for islet transplantation. *N Engl J Med* **355**:1318–1330, 2006.

Shi Q and Xie K. Experimental animal models of pancreatic cancer (review). *Int J Oncol* **17**: 217–225, 2000.

Shi Y, Sahu RP and Srivastava SK. Triphala inhibits both in vitro and in vivo xenograft growth of pancreatic tumor cells by inducing apoptosis. *BMC Cancer* **8**:294, 2008.

Shimabukuro M, Zhou YT, Levi M and Unger RH. Fatty acid-induced beta cell apoptosis: a link between obesity and diabetes. *Proc Natl Acad Sci USA* **95**:2498–2502, 1998.

Singh L, Bakshi DK, Vasishta RK, Arora SK, Majumdar S and Wig JD. Primary culture of pancreatic (human) acinar cells. *Dig Dis Sci* **53**:2569–2575, 2008.

Siveke JT, Lubeseder-Martellato C, Lee M, Mazur PK, Nakhai H, Radtke F and Schmid RM. Notch signaling is required for exocrine regeneration after acute pancreatitis. *Gastroenterology* **134**:544–555, 2008.

Smelt MJ, Faas MM, de Haan BJ and de Vos P. Pancreatic beta-cell purification by altering FAD and NAD(P)H metabolism. *Exp Diabetes Res* **2008**:165360, 2008.

Suemizu H, Monnai M, Ohnishi Y, Ito M, Tamaoki N and Nakamura M. Identification of a key molecular regulator of liver metastasis in human pancreatic carcinoma using a novel quantitative model of metastasis in NOD/SCID/gammacnull (NOG) mice. *Int J Oncol* **31**:741–751, 2007.

Suen PM, Li K, Chan JC and Leung PS. In vivo treatment with glucagon-like peptide 1 promotes the graft function of fetal islet-like cell clusters in transplanted mice. *Int J Biochem Cell Biol* **38**:951–960, 2006.

Suen PM. Isolation, characterization and differentiation of pancreatic progenitor cells from human fetal pancreas. Ph.D. Thesis, The Chinese University of Hong Kong, 2007.

Sutherland AP, Van Belle T, Wurster AL, Suto A, Michaud M, Zhang D, Grusby MJ and von Herrath M. Interleukin-21 is required for the development of type 1 diabetes in NOD mice. *Diabetes* **58**:1144–1155, 2009.

Takahashi N, Kishimoto T, Nemoto T, Kadowaki T and Kasai H. Fusion pore dynamics and insulin granule exocytosis in the pancreatic islet. *Science* **297**:1349–1352, 2002.

Tei E, Mehta S, Tulachan SS, Yew H, Hembree M, Preuett B, Snyder CL, Yamataka A, Miyano T and Gittes GK. Synergistic endocrine induction by GLP-1 and TGF-beta in the developing pancreas. *Pancreas* **31**:138–141, 2005.

Teramura Y, Kaneda Y and Iwata H. Islet-encapsulation in ultra-thin layer-by-layer membranes of poly(vinyl alcohol) anchored to poly(ethylene glycol)-lipids in the cell membrane. *Biomaterials* **28**:4818–4825, 2007.

Vasir B, Aiello LP, Yoon KH, Quickel RR, Bonner-Weir S and Weir GC. Hypoxia induces vascular endothelial growth factor gene and protein expression in cultured rat islet cells. *Diabetes* **47**:1894–1903, 1998.

Verheul HA, Schot LP and Schuurs AH. Effects of sex, gonadectomy and several steroids on the development of insulin-dependent diabetes mellitus in the BB rat. *Clin Exp Immunol* **63**:656–662, 1986.

Wang L, Xie LP and Zhang RQ. Gene expression of gonadotropin-releasing hormone and its receptor in rat pancreatic cancer cell lines. *Endocrine* **14**:325–328, 2001.

Wilkes JJ, Lloyd DJ and Gekakis N. A loss-of-function mutation in myostatin reduces TNF{alpha} production and protects liver against obesity-induced insulin resistance. *Diabetes* **58**: 1133–1143, 2009.

van der Windt DJ, Echeverri GJ, Ijzermans JN and Coopers DK. The choice of anatomical site for islet transplantation. *Cell Transplant* **17**:1005–1014, 2008.

Zekzer D, Wong FS, Ayalon O, Millet I, Altieri M, Shintani S, Solimena M and Sherwin RS. GAD-reactive CD4+ Th1 cells induce diabetes in NOD/SCID mice. *J Clin Invest* **101**:68–73, 1998.

Zhou J, Ou-Yang Q, Li J, Zhou XY, Lin G and Lu GX. Human feeder cells support establishment and definitive endoderm differentiation of human embryonic stem cells. *Stem Cells Dev* **17**:737–749, 2008.

Chapter 8
Current Research of the RAS in Diabetes Mellitus

Type 2 diabetes mellitus (T2DM) accounts for more than 90% of human diabetes and the incidence of new cases is ever increasing worldwide; its prevalence thus places huge demands on healthcare resources and presents a new challenge for the control of disease globally. T2DM is a metabolic disease characterized by hyperglycemia because of enhanced peripheral insulin resistance, impaired β-cell function and decreased β-cell mass. During the development of peripheral insulin resistance leading to an increased demand for insulin secretion, β-cell mass is significantly enhanced by hyperplasia and hypertrophy of β-cells concomitantly with an increase in insulin production as a compensatory mechanism to control normal blood glucose concentrations. As long as β-cell compensatory capacity is maintained, insulin resistance on its own is not sufficient to trigger overt T2DM (Wang & Jin, 2009). In view of this fact, the effective strategy and management of T2DM are to target for the improvement of islet β-cell mass and function rather than peripheral insulin resistance. As of yet, there are still no cures for T2DM and, in view of this fact, basic and clinical research studies are working to elucidate underlying mechanisms and molecules involved in regulating pancreatic islet cell function and T2DM as such findings may produce much needed candidate therapies. This mechanism-driven approach should provide new insights into the development of novel strategies for cost-effective prevention of T2DM and associated metabolic syndrome disorders, and possibly treatment of these conditions as well. There are numerous frontiers in diabetic research dealing with recent developments and new discoveries in the field, including some cutting-edge therapeutic approaches (see articles in special issue by Leung et al., 2006).

Many regulators with potential roles concerning β-cell function and insulin resistance remain to be characterized in detail. Among these, there are a number of novel amino acids, peptides and proteins that may affect pancreatic insulin secretion and insulin sensitivity as well as β-cell growth and differentiation (see articles in special issue by Leung & De Gasparo, 2009). Notwithstanding the existence of these promising candidates, there are three physiological regulators, namely angiotensin II, vitamin D and glucagon-like peptide 1 (GLP-1), which are emerging from the growing list of candidates related to islet function and T2DM. In this Chapter, we will provide a critical appraisal of contemporary research progress on the novel

P.S. Leung, *The Renin-Angiotensin System: Current Research Progress in The Pancreas*, Advances in Experimental Medicine and Biology 690,
DOI 10.1007/978-90-481-9060-7_8,

roles of the renin-angiotensin system (RAS), vitamin D and GLP-1, with particular emphasis on the putative interactions of the pancreatic islet RAS with vitamin D and GLP-1, especially involvement of a RAS-vitamin D-GLP1 axis in the integration of islet function and its clinical relevance to T2DM.

8.1 Basic Studies of the RAS in T2DM

Local RASs play a multitude of autocrine, paracrine and/or intracrine roles in the regulation of specific tissue and organ functions (see review by Paul et al., 2006; also see Chapter 5) compared with circulating RASs (see Chapter 4). As discussed in Chapter 6, local pancreatic RASs are now recognized in various cell types of the pancreas including acinar, ductal, stellate and islet cells. The expression of these pancreatic RASs is modulated in response to various physiological and pathophysiological stimuli including, but name a few, hypoxia, islet transplantation, exposure to high glucose concentrations, diabetes mellitus, pancreatitis, and pancreatic cancer. These functional RASs are proposed to have important endocrine and exocrine functions in the pancreas (see review by Leung, 2007a). Of particular interest in this context is the recently identified local RAS in the isolated mouse pancreatic islets which express all major RAS components including angiotensinogen, an obligatory element of a local RAS (Lau et al., 2004).

8.1.1 RAS Blockade Studies in Animal Models of T2DM

Such islet RAS is upregulated in animal models of T2DM, such as in obesity-induced *db/db* mice (Chu et al., 2006) and in Zucker diabetic fatty (ZDF) rats (Tikellis et al., 2004) and, more importantly, blockade of the RAS improves β-cell mass and function in these two animal models of diabetes. Several additional animal models for the study of islet RAS blockade in T2DM have also been developed recently. For example, long-term (24-week) administration of the angiotensin-converting enzyme (ACE) inhibitor ramipril to diabetic Otsuka Long-Evans Tokushima fatty (OLETF) rats can prevent islet destruction by fibrosis, as evidenced by the expression profile of TGF-β and its downstream signalling molecules (Ko et al., 2004). Similarly, chronic treatment (10 weeks) with either an ACE inhibitor (perindopril) or type 1 angiotensin II receptor (AT1R) blocker (irbesartan) attenuates islet fibrosis and reduces islet cell apoptosis, and oxidative stress in ZDF rats (Tikellis et al., 2004). On the other hand, hyperglycemia, as observed in T2DM, has recently been shown to activate the RAS in pancreatic islet and stellate cells; RAS blockade ameliorates the angiotensin II-induced pancreatic islet inflammation and fibrosis that is exacerbated by chronic exposure to high glucose concentrations (Lupi et al., 2006; Ko et al., 2006).

Consistent with these in vivo findings, exposure of INS-1E cells (a pancreatic β-cell line) to a range of glucose concentrations (0.8, 5.6, 11.1, 28 and 56 mM) induces dose-dependent upregulation of AT_1 receptor transcription and translation (Fig. 8.1a, b). Such AT_1 receptor activation could result in inhibition of glucose-stimulated insulin secretion from the INS-1E cells (Fig. 8.1c). The potential mechanism mediating this effect may be involved in AT_1 receptor activation of NADPH oxidase-induced oxidative stress (Leung & Leung, 2008). The data imply that prolonged exposure to high glucose levels (chronic hyperglycemia) is glucotoxic for β-cells such as INS-1E cells, and concomitantly induces AT_1 receptor upregulation on these cells, thus leading to impaired insulin secretory function. Taken together, these in vivo and in vitro studies provide a convergence of evidence indicating that pancreatic islet RAS activation may be involved in impaired insulin secretion and enhanced oxidative stress-induced islet cell apoptosis and islet fibrosis; in other words, excessive RAS activation may lead to islet dysfunction and decreased β-cell mass.

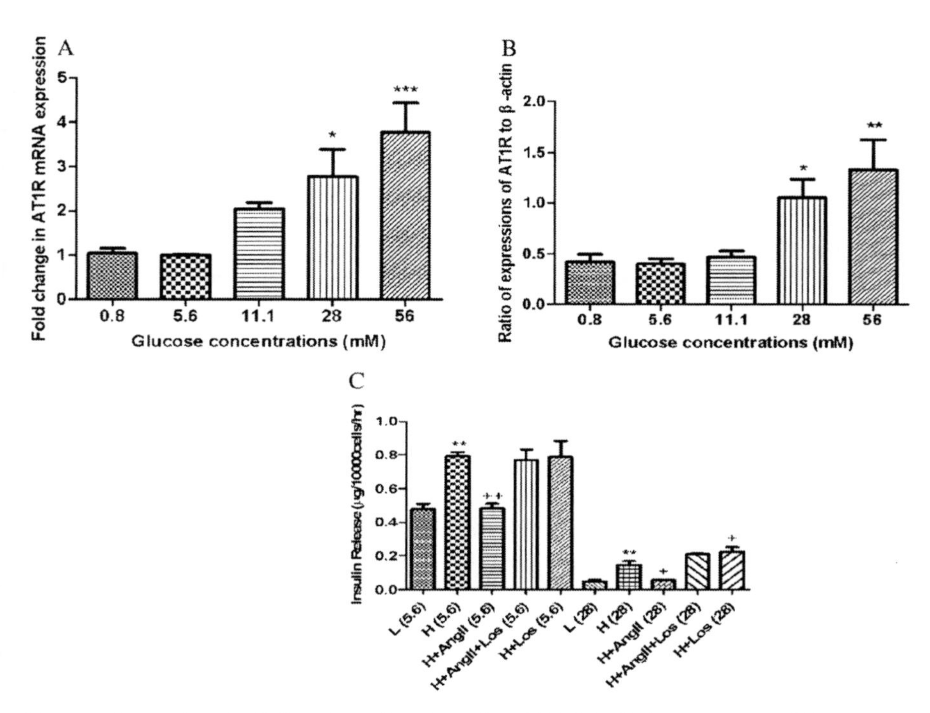

Fig. 8.1 Expression of AT1R and insulin release from INS-1E cells in exposure to different concentrations of glucose. **a** Quantitative PCR analysis. **b** Western blot analysis. **c** Glucose-induced insulin secretion from INS-1E cells in response to 5.6 and 28 mM glucose. All data are expressed as mean±SEM; $^{*}p < 0.05$ vs. 5.6 mM; $^{**}p < 0.01$ vs. 5.6 mM; $+p < 0.05$ vs. cells exposed to 28 mM glucose; $++p < 0.01$ vs. cells exposed to 28 mM glucose (data are extracted from Leung & Leung, 2008)

8.1.2 RAS Blockade-Induced Protective Mechanism

In spite of the aforementioned demonstrations, the precise mechanism(s) by which the local pancreatic RAS modulates islet function remains to be elucidated. The hypothesis that there is a change in AT_1 receptor expression in T2DM which enables endogenous levels of angiotensin II to impair islet function has recently been tested in obesity-induced *db/db* mice (Chu et al., 2006). Though the pancreatic islet AT_1 receptor had no obvious effects on normal islet function, it was upregulated in the obesity-induced *db/db* T2DM mice; this upregulation has negative effects on glucose-stimulated islet insulin secretion and (pro)insulin biosynthesis and on islet blood flow (Chu et al., 2006). These findings provide, at least partly, a novel explanation for the reduced incidence of T2DM that has been observed in a number of clinical trials of AT_1 receptor antagonism in individuals at high risk for the disease (Chu et al., 2006; Leung & De Gasparo, 2006). In addition, AT_1 receptor antagonism (for 8 weeks) was shown to attenuate NADPH oxidase-induced oxidative stress in isolated islets from db/db mice in vivo; this in turn results in a down-regulation of uncoupling protein 2 (UCP2) expression, which is associated with augmented β-cell insulin secretion and reduced loss of β-cells to apoptosis (Chu & Leung, 2007). In keeping with these findings, chronic AT_1 receptor antagonism has been shown to improve islet cell function and structure in OLETF rats and *db/db* mice, an effect that is apparently mediated by NADPH oxidase-induced oxidative stress (Nakayama et al., 2005; Shao et al., 2006). In light of the above evidence, it is plausible to propose that AT_1 receptor activation in T2DM mediates increases in oxidative stress followed by resultant reductions in UCP2 activity which, in turn, leads to pancreatic islet β-cell dysfunction. We have recently shown that angiotensin II can exert a glucose-induced action on Kv currents in isolated mouse islets. In that study, exogenous addition of angiotensin II reduced Kv current amplitude under normal, but not high, glucose conditions. This angiotensin II-induced effect on Kv channel activity was abolished by inhibition of angiotensin II type 2 receptor (AT_2 receptor). Our data suggest that hyperglycemia alters β-cell function via mediation of the Kv channel expression and activity, and that this process may be associated with AT_2 receptor, but not AT_1 receptor (Chu et al., 2010). Nevertheless, the precise mechanism(s) whereby the AT_2 receptor modulates channel activity, and thus pancreatic β-cell secretory function, has yet to be fully determined.

In summary, basic research studies have garnered convergent data indicating that islet RAS has a novel role in islet function and that its dysregulation may lead to impaired islet function and structure, and ultimately T2DM. In this context, the islet RAS regulates pancreatic islet blood flow, oxygen tension, and islet (pro)insulin biosynthesis (Lau et al., 2004; Kampf et al., 2005; Chu et al., 2006). Meanwhile, it also mediates NADPH oxidase-driven UCP2 activation and thus oxidative stress-induced β-cell apoptosis and fibrosis (Chu & Leung, 2007). Blockade of the islet RAS improves β-cell secretory function, islet oxidative damage and structure, and glucose tolerance in experimental T2DM models (Chu et al., 2006; Chu & Leung, 2007; Cheng et al., 2009). These findings indicate that inhibition of islet RAS activation is an alternative approach to protecting islet cell function and preventing the

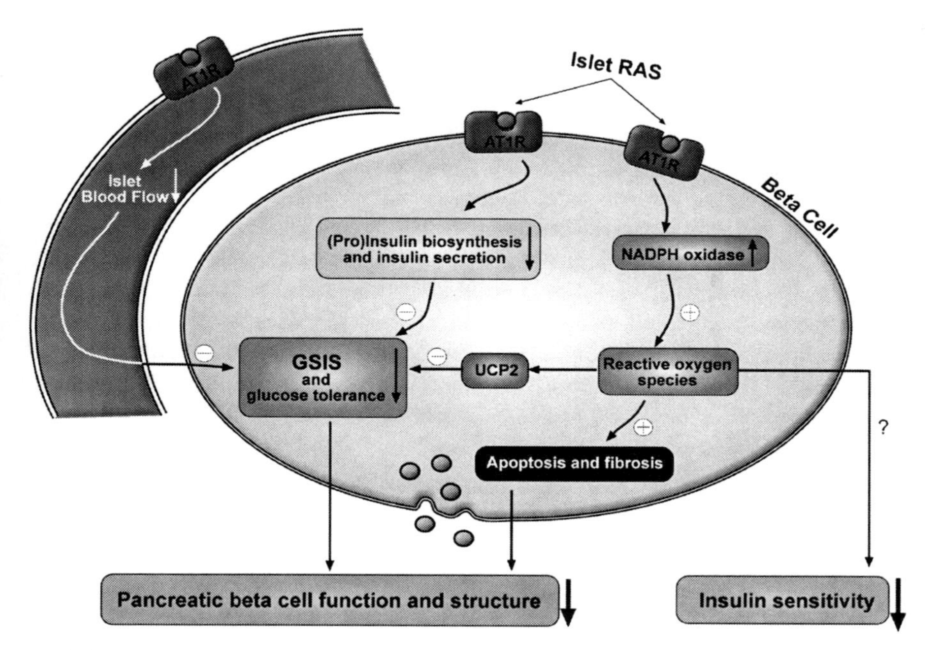

Fig. 8.2 Summarization of the currently proposed mechanism(s) by which AT1R mediates islet cell function and structure as well as insulin resistance in T2DM

development of overt T2DM (Leung, 2007b). Furthermore, RAS upregulation associated increases in islet oxidative stress and/or reactive oxygen species (ROS) may induce pancreatic β-cell dysfunction and increase insulin resistance in peripheral tissues, thus contributing to development of cardiometabolic syndrome abnormalities (Leung & De Gasparo, 2006). Figure 8.2 summarizes the currently proposed mechanism(s) by which the AT_1 receptor may mediate islet cell function and structure as well as insulin resistance.

8.2 Clinical Studies of the RAS in T2DM

8.2.1 Recent Clinical Trials on RAS Blockers

Complementary to the basic experimental studies, a host of large clinical trials in patients at high risk for T2DM have reported that RAS blockade delays and/or prevents the new onset of T2DM incidence (Scheen, 2004a, b). These clinical trials include, but are not limited to, Heart Outcomes Prevention Evaluation (HOPE), Losartan for Interventions for Endpoints in Hypertension (LIFE), Nateglimide And Valsartan in Impaired Glucose Tolerance Outcomes Research (NAVIGATOR), Captopril Prevention Project (CAPP), (Prevention of Events with Angiotensin Converting Enzyme inhibition (PEACE), and Renal Outcomes With Telmisartan,

Ramipril or Both in People at High Vascular Risk (ONTARGET). The conclusion that RAS blockade induces T2DM protective effects is further supported by a meta-analysis of randomized clinical studies (Abuissa et al., 2005; Elliott & Meyer, 2007a, b). For example, a meta-analysis of hypertension treatment trials (Elliott & Meyer, 2007a, b) has suggested that angiotensin receptor blockers (ARBs) are more effective than ACE inhibitors (ACEIs) in reducing the risk of T2DM (Lam & Owen, 2007). Among the clinical trial studies, the PEACE Trial Investigators (2008) study of 8290 patients over a period of 4.8 years in Australia showed that RAS blockade with trandolapril (in subjects with stable coronary disease, preserved cardiac function and virtually normal blood pressure) was associated with a 17% reduction in the incidence of T2DM (HR 0.83 (0.72, 0.96; $p = 0.01$)), though there was no risk reduction for cardiovascular endpoints and this finding appears to be robust. More recently, the ONTARGET investigators (2008) reported that they found that ACE inhibition (ramipril in 1,412 subjects) and angiotensin receptor blockade (Telmisartan in 1,423 subjects) reduced T2DM risk after 4.8 years to 6.7 and 7.5%, respectively, which is substantially lower than the 34% risk expected from comparable HOPE study data (Yusef et al., 2000). And there is further specific pathophysiological evidence for reduction of insulin resistance and for improvement in insulin secretion with RAS blockade in humans. In view of these substantial clinical data, it is plausible to conclude that blockade of the RAS, in particular with ARBs, may be an efficacious clinical intervention for prevention of T2DM (Kintscher et al., 2008).

8.2.2 DREAM Trial and Its Implications

The DREAM trial recently reported that 3 years of treatment with an ACE inhibitor (ramipril) caused an increase in the regression from impaired to normal glucose tolerance ($p = 0.01$; DREAM Trial Investigators, 2006). However, contrary to the HOPE study, they did not find evidence of a reduction in T2DM risk. These negative findings may be attributable to study limitations, in particular the premature termination of the study after 3 rather than 6 years, The DREAM trial was stopped early because the rosiglitazone arm had shown reduction in T2DM risk. Alternatively, the negative findings may be related to the treatment of normotensive rather than hypertensive subjects and the assumption that the two treatment arms of the planned crossover study would work in comparable ways. Hence the subjects were less RAS active than subjects in trials with hypertensive patients, thus suggesting that longer periods of RAS blockade would be indispensable for detection of reduced T2DM risk in DREAM subjects. Regardless, the observation that ramipril did not ameliorate T2DM in the DREAM study did cast some doubt on the underlying mechanism(s) by which RAS blockade is protective against T2DM. Importantly, however, the parallel ramipril arm in this 2×2 factorial study had shown improvement in impaired glucose tolerance (IGT)/impaired fasting glycemia (IFG) regression at 3 years ($p = 0.01$). IGT is generally accepted to carry a 25% risk of progression to T2DM over 5 years in white subjects, thus suggesting that RAS blockade with ramipril could therefore be expected to reduce T2DM risk in the long term. Other factors likely to have reduced the power of the shortened ramipril arm of the DREAM study to detect reductions in T2DM risk include the absence of hypertension or of advanced pre-diabetes in recruited subjects. It is important note that the benefits of RAS blockers and thiazolidones may not be simply additive, thus further under powering the ACE inhibitor arm of the 2×2 factorial DREAM study. Finally, the fact that the thiazolidone treatment (rosiglitazone) was associated with increased risk of heart failure in the DREAM study means that this agent is unlikely to be suitable for use at the population level. The evidence to date, therefore, suggests that there is currently a need (i) to establish whether there are mechanisms by which RAS blockade could reasonably be expected to reduce T2DM risks as evidenced by proof-of-concept studies, and (ii) to determine the value of RAS blockade for T2DM risk reduction in healthy subjects.

In light of the basic research and clinical data discussed above, it is reasonable to postulate that the protective effects against glycemia induced by RAS blockade involve the local pancreatic islet RAS. The proposed mechanisms whereby RAS inhibition is protective against T2DM include improved pancreatic islet cell function and structure mediated through enhancement of islet blood flow and (pro)insulin biosynthesis with concomitant reductions in ROS-induced apoptosis and fibrosis and reduced peripheral tissue insulin sensitivity mediated through an increase in blood flow of skeletal muscle.

8.3 Current Research on the RAS-Vitamin D-T2DM Axis

Vitamin D is best known for its involvement in the maintenance of normal calcium balance. However, it has other functions, and one of these is the recently proposed notion that normal insulin release from the pancreas needs vitamin D in addition to a sufficient supply of calcium. Therefore, pancreas malfunction, such as occurs in T2DM, may have originate partly as a result of vitamin D deficiency, which is a fairly common problem. In support of this view are experimental and clinical data showing that vitamin D deficiency impairs insulin secretion and increases insulin resistance, two typical markers of increased risks of developing T2DM.

8.3.1 Vitamin D and RAS in T2DM

Recent data have shown a link between vitamin D and negative RAS regulation. In this context, identification of receptors for $1\alpha,25(OH)_2D_3$ (1,25-dihydroxyvitamin D_3, the VDR) and of 1α-OHase vitamin D activating enzyme in many tissues (e.g. immune cells, the vasculature, pancreatic islet β-cells) has led to the recognition of many novel roles for vitamin D, in addition to its classical roles in calcium and bone homeostasis (Holick, 2003; Mathieu et al., 1994; Hewison et al., 2007).

$1\alpha,25(OH)_2D_3$ binding to the nuclear VDR and to VDR in plasma membrane caveolae generates slow genomic and rapid non-genomic responses, respectively (Norman, 2006). In particular, stimulation of islet insulin secretion by activated vitamin D is largely a rapid non-genomic effect following activation of the caveolae-associated cell membrane VDR by 1,25-dihydroxyvitamin D (Kajikawa et al., 1999). Interestingly, a review of the literature shows that islet β-cell insulin secretion is reduced in hypovitaminosis D and increases, sometimes to normal, with replacement of vitamin D in whole animals, isolated islets, and humans (Mathieu et al., 2005; Bourlon et al., 1999; Boucher et al., 1995). First and second phase of glucose-stimulated islet insulin secretion, impaired in early and sustained vitamin D deficiency respectively, are restored by adequate dietary supplementation with vitamin D, in vitro and in vivo (Kadowski & Norman, 1984; Cade & Norman, 1986). Figure 8.3 depicts a schematic representation of the mechanism by which vitamin D interacts with caveolae-bound and nucleus-bound VDR mediated transduction pathways in the regulation of pancreatic β-cell function.

In keeping with these findings, clinical studies have shown an increased prevalence of T2DM in association with vitamin D-deficiency in humans (Scragg, 1981;

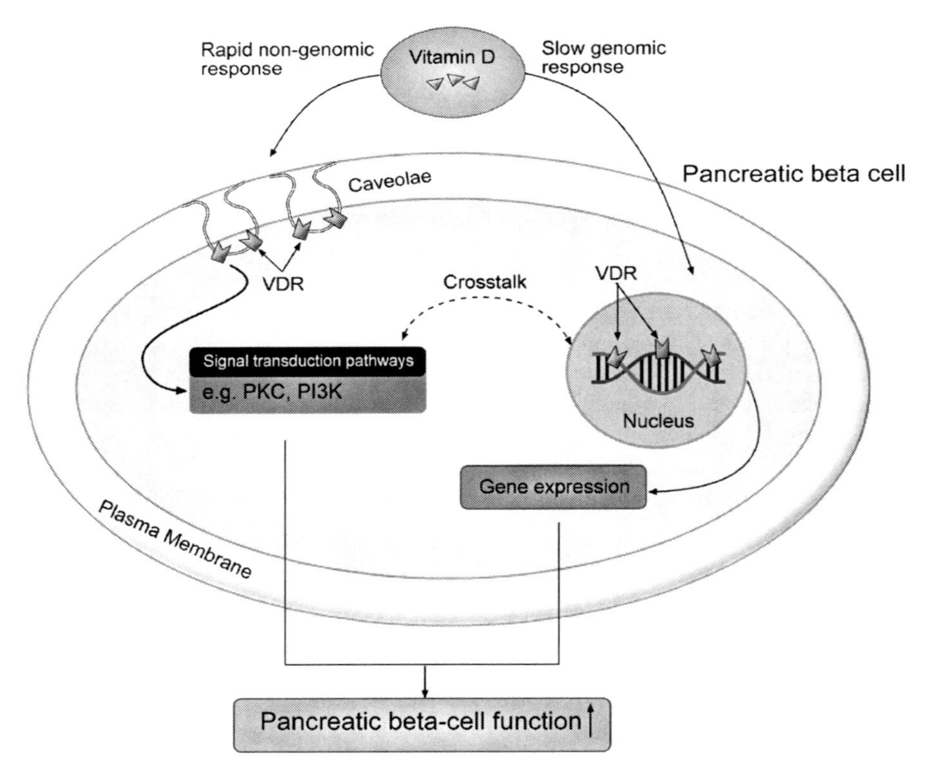

Fig. 8.3 A schematic representation of "rapid" non-genomic and "slow" genomic responses mediated by VDR located on the caveola and cell nucleus in the pancreatic beta cells

Scragg et al., 1990, 2004; Forman et al., 2007). Hypovitaminosis D is also associated with metabolic syndrome defining abnormalities and with increased risk of both T2DM and cardiovascular disease (Boucher, 1998; Chiu et al., 2004; Mathieu et al., 2005). Additional supportive data show an association of reduced ultraviolet-B availability with increased risk of T2DM, possibly attributable to altered immunobiology (Staples et al., 2003). Nevertheless, the mechanisms by which vitamin D contributes to the maintenance of normal islet cell physiology remain obscure. Proposed pathways include increasing intracellular calcium ion concentration, which is necessary for activation of some proteases that release insulin from (pro)insulin (Davidson et al., 1988; Sergeev & Rhoten, 1995). Furthermore, vitamin D status has direct associations with insulin secretory function in humans (Clark et al., 1987; Ishida & Norman, 1988; Billaudel et al., 1989; Lee et al., 1994; Boucher et al., 1995; Bland et al., 2004), which can be explained by the fact that circulating 25-hydroxyvitamin D_3 is activated to form $1\alpha,25(OH)_2D_3$ in the islets, which themselves express the VDR.

Hypovitaminosis D is associated with elevated blood pressure in humans (Scragg et al., 2007; Chiu et al., 2004) and vitamin D has been shown, experimentally, to reduce renal RAS activity through suppression of renin secretion, specifically (Li et al., 2004), providing a likely mechanism for blood pressure reductions reported with vitamin D supplementation in humans, especially those with hypovitaminosis D (Lind et al., 1988, 1989). Relevant to these effects, it is noteworthy that RAS activity is several folds greater in VDR-null mice than wild-type mice (Li et al., 2002, 2004; Kong & Li, 2003). Furthermore, there is substantial data from both basic and clinical studies demonstrating that increases in pancreatic RAS activity reduce islet β-cell function and structure, whilst increasing the activity of pro-inflammatory factor genes and proteins in experimental animal models of T2DM.

8.3.2 Role of Vitamin D in Modulating Islet RAS Expression and Function

In view of this convergence of evidence, we hypothesize that vitamin D regulates pancreatic islet function (insulin secretion) indirectly by modulating pancreatic islet RAS expression and activity, thereby supporting β-cell function. We are now testing this hypothesis in both in vitro and in vivo models of vitamin D deficiency and repletion and in T2DM; this work should provide a proof-of-principle that vitamin D modulates islet RAS activity as well as reducing circulating RAS activity, which in turn regulates islet β-cell function. Interestingly, we have collected data (unpublished) showing that the expression of major RAS components in isolated pancreatic islets were upregulated under a high glucose condition (28 mM) compared to that under a low glucose condition (5.6 mM). For example, AT_1 receptor translation and transcription were markedly increased in the high glucose condition while co-exposure of the islets to active vitamin D (calcitriol, most notably at 10^{-9} M) down-regulated AT_1 receptor expression under high glucose concentration

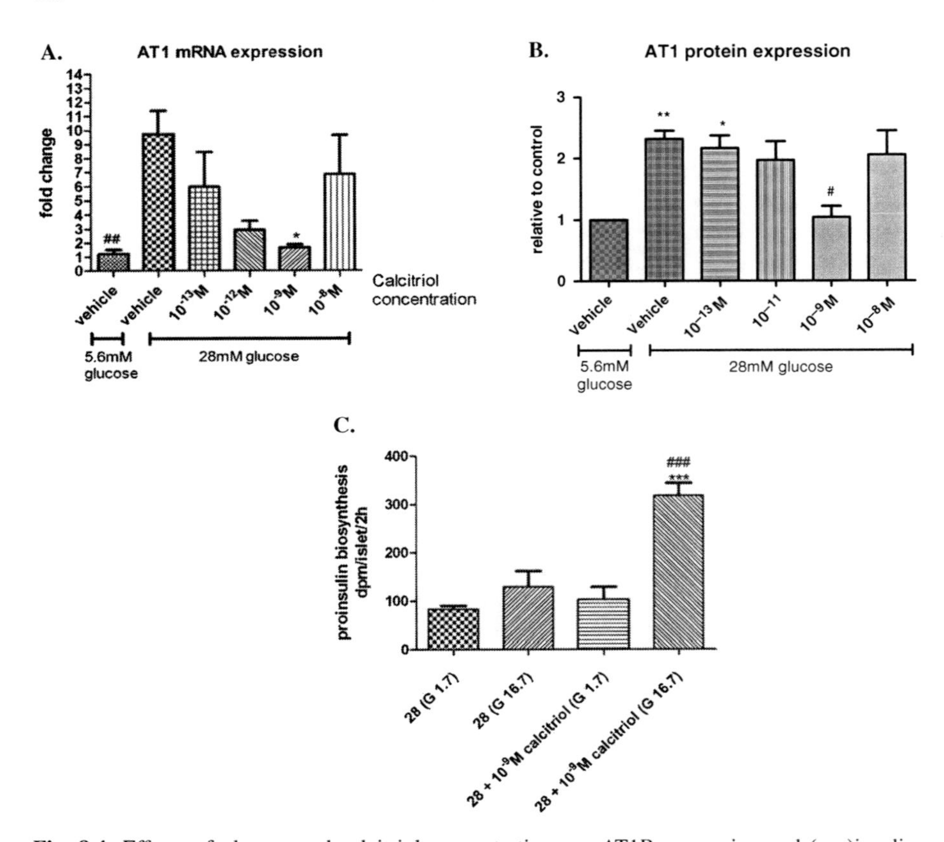

Fig. 8.4 Effects of glucose and calcitriol concentrations on AT1R expression and (pro)insulin biosynthesis in isolated islets. **a** Real-time PCR analysis of AT1R. **b** Western blot analysis of AT1R. **c** (Pro)insulin biosynthesis. $^{*}p<0.05$ vs. 28 mM glucose + vehicle; $^{\#}p<0.05$ vs. 28 mM glucose + vehicle: $^{\#\#}p<0.01$ vs. 28 mM G + vehicle (Data are expressed as means ± SEM, n=6–8 in each group) (Cheng & Leung, unpublished data)

(Fig. 8.4). On the other hand, (pro)insulin biosynthesis in isolated islet incubates was significantly enhanced in the presence of calcitriol at high but not low concentrations of glucose (Fig. 8.4). Our preliminary data suggest that vitamin D may exert beneficial actions on pancreatic islet function by down-regulating expression of RAS components, such as the AT_1 receptor, in a high glucose and/or hyperglycemic state such as that observed in patients with T2DM. Notwithstanding relevant findings from our laboratory and others supporting the modulation by vitamin D of islet RAS in T2DM, the existence of a vitamin D-RAS-T2DM axis, the mechanism(s) by which hypovitaminosis D-dependent RAS activation reduces islet β-cell function thus T2DM warrant further investigation.

8.3.3 Expression of Islet RAS in VDR Knockout Mice

As discussed in previous section, vitamin D contributes to regulating various physiological functions. Many clinical evidences have substantiated intricate relationship between vitamin D levels and cardiovascular diseases, and individuals with reduced vitamin D productions induced by such conditions as insufficient exposure to UVB, aging and work shift at night, are associated with abnormal blood pressure (Krause et al., 1998; Hypponen et al., 2008); meanwhile, high vitamin D levels can attenuate blood pressure in people with risks of hypovitaminosis D (Scragg et al., 2007). Consistently, several animal studies have also shown that vitamin D deficient rats have significant impact in cardiovascular functions (Weishaar & Simpson, 1987). As discussed, the RAS is a critical regulator in cardiovascular function and thus it is plausible to propose a close interaction between vitamin D and islet RAS. In order to address the novel role of vitamin in modulating the islet RAS, VDR knockout mice are indispensable approach to test our hypothesis.

To address this issue, several animal models of VDR ablation mimicking the conditions of vitamin D deficiency or resistance are available. For example, mouse model of VDR ablation is based on the genetic 1,25-dihydroxyvitamin D resistant rickets; different mutations are characterized in the DNA-binding domain (DBD) and ligand-binding (LBD) domain of the VDR (Malloy et al., 1997). In this regard, VDR knockout mice are generated by targeted gene disruption, in particular with a targeted ablation of exon 1, 2 and/or 3 of the mice VDR (Li et al., 1997; Yoshizawa et al., 1997; Van Cromphaut et al., 2001; Erben et al., 2002). Generally, a targeting plasmid is constructed by replacing the exons of interest encoding the specific zinc finger of the receptor DBD critical for the function of the VDR; the targeting vector is then linearized and introduced into embryonic stem cells (ES cells) by electroporation, where the homologous pieces of DNA recombined while original DNA is replaced by new DNA in the vector. The growth of ES cells in selective medium allows the identification of the homologous recombination and screened ES cells are then injected into 3.5-day old mouse blastocytes and transferred into uterus of female mice. Chimera mice are derived from two independent ES cells clones indicative of transmission of the germ line. After genotype tests and confirmation, the VDR ablation mice are successfully constructed (Li et al., 1997). Figure 8.5 summarizes with a schematic representation how VDR mice are generated. Apart from VDR, knockout mice with dysfunction of vitamin D metabolic enzymes are also useful tools to study vitamin D function in particular with its metabolic pathways such as CYP27A1, a vitamin D 25-hydroxylase (Repa et al., 2000) or CYP24, a 25-hydroxyvitamin D-24-hydroxylase (St-Arnaud et al., 2000), and 25-hydroxyvitamin D-1α-hydroxylase (Dardenne et al., 2001; Panda et al., 2001). The phenotype of these mice is rescued by $1,25(OH)_2D_3$ treatment, a high-calcium and phosphorus diet therapy.

Since the key role of RAS in cardiovascular function and its contribution to the development of hypertension (Connell et al., 2008), the linkage between RAS and

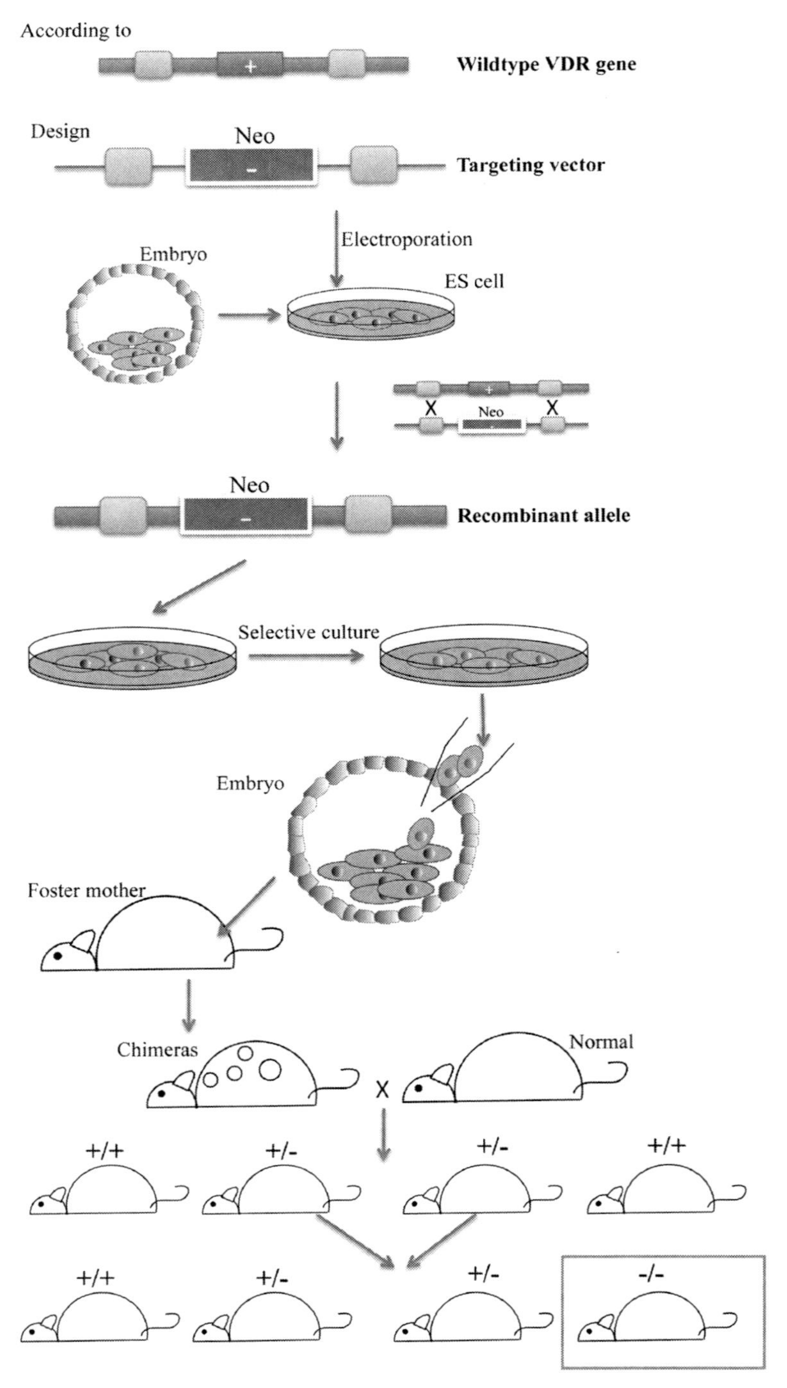

Fig. 8.5 A schematic diagram showing the construction of VDR knockout mice. Customized genetic changes are introduced into ES cells where the desired modification is selected and

vitamin D has begun to draw attention. The pioneering work was performed by YC Li's laboratory at The University of Chicago, showing an increase in kidney renin expression from VDR knockout mice; this effect might be due to the lack of vitamin D-mediated reduced activity of the renin promoter (Li et al., 2002). In additional to renin, several other RAS components were also regulated by VDR activation thus affecting its downstream functions (Freundlich, 2008). Against this background, we propose that vitamin D knockout has modulatory actions on the expression and function of the islet RAS and thus islet cell and T2DM. In collaboration with YC Li, we examine the expression of RAS components using isolated islets from VDR knockout and wild-type mice. Our preliminary data have shown that protein expression of several islet RAS components including renin and AT2R are subject to regulation compared with those islets from wild-type mice (Fig. 8.6). Moreover, the mRNA expression of renin receptor, AT_1 receptor and ACE are also regulated in VDR knockout mice compared with wild-type mice islet (Fig. 8.6). In conclusion, our preliminary results suggest a potential interaction of vitamin D and islet RAS and thus its relevance to T2DM. Figure 8.7 is a highly schematic presentation proposing how vitamin D modulates the circulating, renal and islet RAS and thus affecting their respective functions.

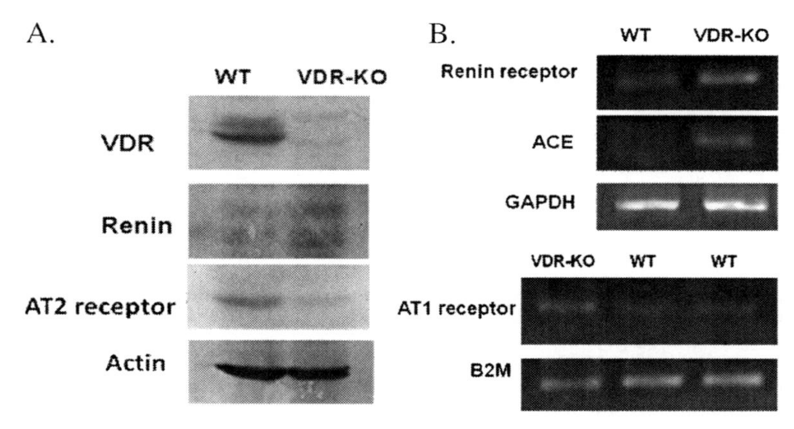

Fig. 8.6 Expression of VDR and RAS components in islets isolated from VDR knockout (KO) and wild-type (WT) mice. **a** protein expression of VDR, renin and AT_2 receptor by Western blot analysis. **b** Gene expression of renin receptor, ACE and AT_1 receptor mRNA by RT-PCR analysis (Cheng & Leung, unpublished data)

Fig. 8.5 (continued) enriched. ES cells are microinjected into mouse blastocysts and then are injected into foster mothers. The pups born from those embryos called chimeric are mated with normal mice and heterozygous offspring is finally bred to transmit the ablation gene to their progeny. Mice in the *blue frame* are homozygous or knockout mice. For interpretation of the references to colour in this figure legend, please be referred to the online version

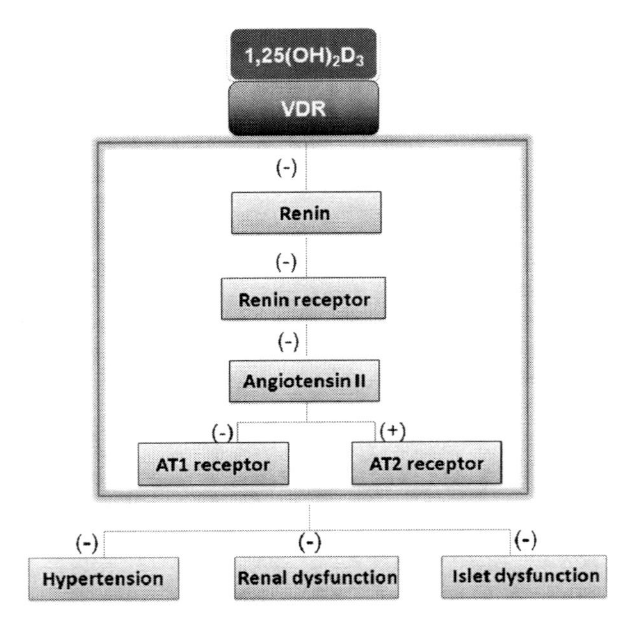

Fig. 8.7 Proposed role of vitamin D in regulating blood pressure, renal function and islet function via its modulatory actions on circulating RAS, renal RAS and islet RAS). (–), decrease or prevent; (+), increase or permit

8.4 Current Research on the RAS-GLP1-T2DM Axis

Incretins are insulinotropic peptide hormones that are released from intestinal endocrine cells in response to glucose load; that is, they enhance insulin secretion from pancreatic β-cells. There are two main incretin peptides, namely glucose-dependent insulinotropic peptide or gastric inhibitory peptide (GIP) and glucagon-like peptide-1 (GLP-1), which is produced in enteroendocrine K and L cells, respectively. GLP-1 originates from a post-translational cleavage of the glucagon gene by the prohormone convertase and, in human, occurs in two active forms, namely GLP-1(7–36) and GLP-1 (7–37). The action of GLP-1 is mediated via activation of the membrane-bound GLP-1 receptor (GLP-1R), a specific seven transmembrane receptor G-protein coupled receptor, which was first functionally expressed and identified in rat (Thorens, 1992) and later in human pancreatic cells (Thorens et al., 1993). Activation of GLP-1R is generally concomitant with stimulation of various second messenger pathways in β-cells, such as cAMP, protein kinase A (PKA), cAMP regulated guanine nucleotide exchange factors (cAMPGEF or Epac), calcium/calmodulin, and mitogen-activated protein kinase (MAPK)/phosphatidylinositol-3-OH (PI-3) kinase related pathways; activation of these second messenger pathways mediates the acute effects of GLP-1 on glucose sensing and insulin secretion, as well as GLP-1's chronic effects on insulin biosynthesis and secretion (see review by Doyle & Egan, 2007).

8.4.1 Role of DPP-IV Inhibition in T2DM

A growing body of evidence has shown GLP-1 to be a promising anti-diabetic agent (Holst, 2002). GLP-1 stimulates insulin secretion in a glucose-dependent manner, suppresses glucagon release (Ahren, 1998), stimulates non-insulin-dependent glucose uptake (Prigeon et al., 2003), and delays gastric emptying (Naslund et al., 1999). However, the usefulness of native GLP-1 in the clinic is limited by its short plasma half-life of approximately 2 min and its peptidic nature (Zander et al., 2002). It is degraded rapidly in serum from its N-terminus by dipeptidyl peptidase-IV (DPP-IV) (Kieffer et al., 1995). DPP-IV, present in the blood and on the cell membranes, cleaves active GLP-1 (7–36) to yield the inactive GLP-1 (9–36). Modifications have been made to GLP-1 to enhance its biological half-life and thus its efficacy in vivo; exendin-4 is a thus enhanced GLP-1R agonist now available for T2DM therapy (Wang & Brubaker, 2002). Exendin-4 is not a substrate for DPP-IV, has a half-life of 4–5 h, and has a greater potency than native GLP-1. Meanwhile, direct DPP-IV inhibition is an alternative approach to improving pancreatic islet function and structure. In fact, orally available inhibitors of DPP-IV have been developed and are available for therapeutic use in patients with T2DM. A number of in vitro and in vivo studies support the view that DPP-IV inhibition improves β-cell function by enhancing insulin secretion and augmenting β-cell proliferation (Burkey et al., 2005; Green et al., 2005). Consistent with the experimental animal studies, human data also indicate that DPP-IV inhibition contributes to better glycemic control (Ahren et al., 2004). However, DPP-IV inhibitors are not a very specific, acting on DPP-IIX and DPP-IX in addition to DPP-IV. Hence administration of a DPP-IV inhibitor may affect the degradation of many peptides. Thus, it would be preferable for there to be a DPP-IV inhibitor that retains its superior potency, but with better selectivity for DPP-IV.

8.4.2 Interaction of Islet RAS and GLP-1

As discussed in previous sections, a local pancreatic islet RAS exists and plays a role in glucose-stimulated insulin secretion in the pancreas (Lau et al., 2004); islet RAS is upregulated in animal models T2DM (Chu et al., 2006). Additionally, blockade of this islet RAS improves islet β-cell function and glucose tolerance, and reduces ROS-mediated pro-inflammatory factor gene and protein production, thereby confirming RAS's clinical relevance to T2DM and insulin resistance (Chu et al., 2006; Chu et al., 2007; Leung 2007a, b; Leung & De Gasparo, 2006). Meanwhile, both GLP-1 and RAS have common target effects on pancreatic islet function and structure as well as insulin resistance, acting through different pathways in muscle, adipose, liver and, in particular, the pancreas (Leung & de Gasparo, 2006). Against this background, we propose that there is an interaction between the RAS and GLP-1 in the pancreas, i.e. the existence of an RAS-GLP1-T2DM axis—which could be influenced to exhibit beneficial effects on T2DM. We thus hypothesize that combined use of a DPP-IV inhibitor and an AT_1 receptor antagonist would have some degree of additive, and perhaps even synergistic, effects toward improving pancreatic islet structure and function as well as insulin resistance. Examination of this hypothesis first requires proof-of-principle studies to test whether a workable and mechanism-driven combination therapeutic approach may be more efficacious than either treatment alone and multifunctional in the treatment and management of T2DM.

In order to test this hypothesis, we performed in vitro and in vivo studies of pancreatic islet secretion and morphology using obesity-induced T2DM (*db/db*) mice. To assess the in vitro effects, isolated islets from the diabetic mice were incubated with the AT_1 receptor antagonist valsartan (1 μM) and/or the GLP-1 receptor agonist exendin-4 (100 nM), and followed with measurement of glucose-stimulated insulin secretion. As expected, insulin release from *db/db* islets was reduced relative to control mice. Exendin-4 or valsartan mono-therapy doubled insulin release from *db/db* islets relative to controls ($p < 0.05$), while combination treatment of both drugs enhanced insulin release to a level comparable with normal islets (Fig. 8.8). Meanwhile, in vivo studies showed that after treatment with the DPP-IV inhibitor LAF237 (1 mg/kg/day) and/or valsartan (10 mg/kg/day) for 8 weeks, β-cell mass, islet proliferation, and islet apoptosis were significantly increased in *db/db* islets of both the mono-therapy and combination treatment groups, relative to their respective controls, with particularly pronounced improvements being observed in animals subjected to combination treatment (Fig. 8.8). Decreased expression of fibronectin and collagen-1 (markers of fibrosis) as well as of dihydroethidium (a marker of superoxide) in the treated groups suggests that the treatments reduced islet fibrosis and ROS production. These in vitro and in vivo data indicate that combination LAF237 + valsartan treatment has additive beneficial effects on pancreatic β-cell structure and function relative to the mono-therapeutic effects (Cheng et al., 2008). Nevertheless the caveat that DPP-IV inhibitors are not highly selective remains a concern for their long-term use. In particular, in addition to their effects on GLP-1, DPP-IV inhibitors can also influence GIP and putitary adenylate cyclase activating polypeptide (PACAP), the latter of which affects hormones such as peptide YY and substance P (Mest & Mentlein, 2005). Some of these peptides, such as substance P, may influence RAS activities in some tissues (Diz et al., 2002), which may then affect the combination effect of the DPP-IV inhibitor and AT_1 receptor blocker.

8.4.3 Potential Role of Vitamin D in Modulating Islet RAS and GLP-1

While the aforementioned findings, among others, indicate that RAS, vitamin D and GLP-1 may regulate islet cell function, and thus influence T2DM pathology, the putative mechanism(s) of this regulation remain ambiguous. The discussions in this chapter, however, provide a basis for ongoing and future lines of research examining the potential roles of this threesome of factors that may be involved in regulating the development of T2DM. Elucidation of these interactions will advance knowledge in

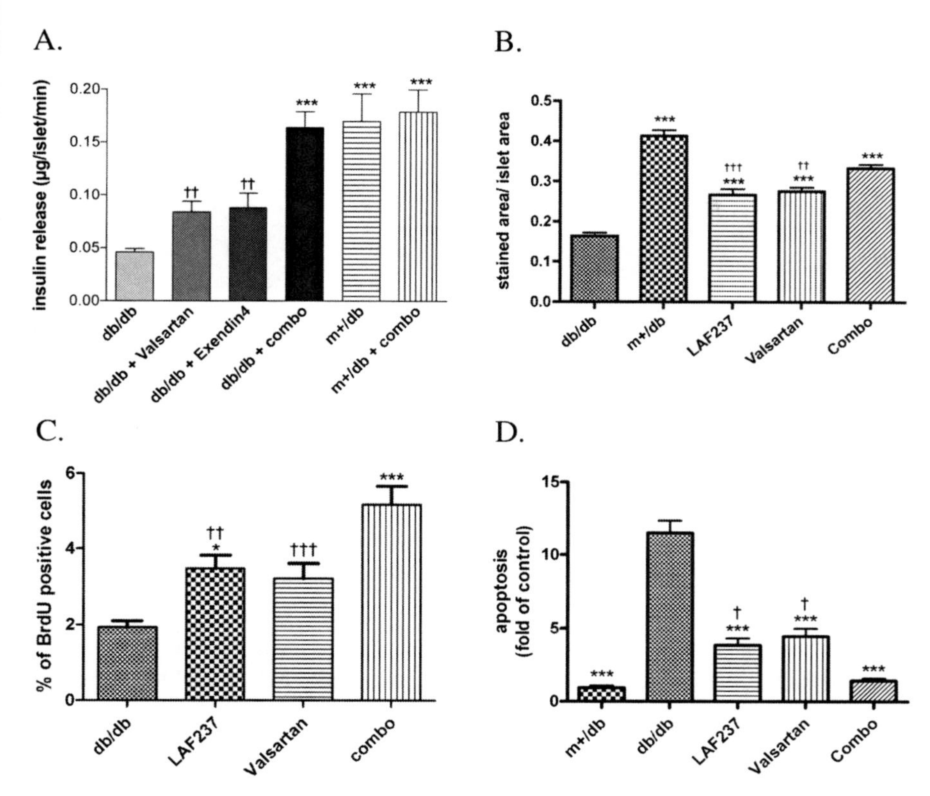

Fig. 8.8 In vitro and in vivo studies of exendin-4, LAF237 and valsartan on islet function from isolated islets of diabetic (db/db) and m+/db (non-diabetic) mice. **a** Effects of exendin-4 and valsartan on islet insulin release. **b** Effects of LAF237 and valsartan on beta-cell mass. **c** Effects of LAF237 and valsartan on islet proliferation. **d** Effects of LAF237 and valsartan on islet apoptosis. $^{***}p < 0.001$ vs. H_2O treated db/db groups; $^{\dagger\dagger}p < 0.01$, $^{\dagger\dagger\dagger}p < 0.001$ vs. combination-treated db/db groups (Combo). All data are expressed as mean ± SEM (10 islets/mice, 5–8 mice/group) (data are extracted from Cheng et al., 2008)

diabetes research. It is indeed intriguing that there have been demonstrations of "cross talk" between the RAS-vitamin D and the RAS-GLP-1 pathways, which have common targets in the pancreas and other peripheral tissues, and that these demonstrations further suggest that this cross talk may be harnessed to influence islet function and β-cell mass as well insulin sensitivity. More importantly, pharmacological use of this kind of combination therapy (i.e., AT_1 receptor antagonist + vitamin D or AT_1 receptor antagonist + DPP-IV inhibitor) undoubtedly provides a promising and mechanism-driven approach for preventing and treating T2DM and metabolic syndrome. In addition, once the full findings from these basic studies are available, it would be appropriate to pursue an adequately powered randomized trials with the following aims: (1) determine whether Vitamin D supplementation in human hypovitaminosis D reduces RAS activity; (2) determine whether certain

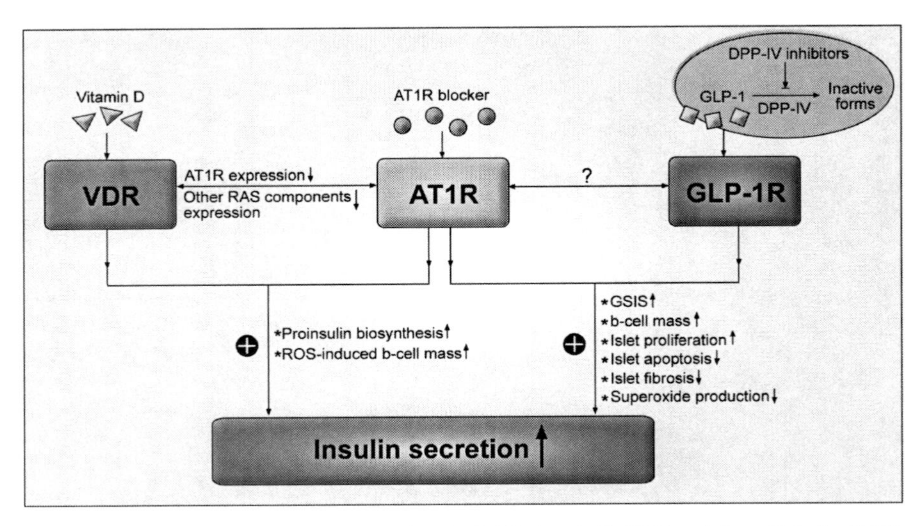

Fig. 8.9 A highly schematic model proposing the RAS-vitamin D-GLP1 axis by which the AT1R interacts with VDR and GLP-1R in the regulation of islet cell mass and insulin secretory function

vitamin D analogues have effects on RAS activity that are superior to those of natural vitamin D_3; and (3) determine whether the combination of vitamin D with RAS blockade may have summative benefits in hypovitaminosis D. Figure 8.9 provides a summary of the proposed mechanism(s) of the potential interactions of the RAS, vitamin D and GLP-1 involved in modulating islet cell function and thus T2DM.

References

Abuissa H, Jones PG, Marso SP and O'Keefe JH. Angiotensin-converting enzyme inhibitors or angiotensin receptor blockers for prevention of type 2 diabetes: a meta-analysis of randomized clinical trials. *J Am Coll Cardiol* **46**:821–826, 2005.

Ahren B, Landin-Olsson, M, Jansson PA, Svensson, M, Holmes D and Schweizer A. Inhibition of dipeptidyl peptidase-4 reduces glycemia, sustains insulin levels, and reduces glucagon levels in type 2 diabetes. *J Clin Endocrinol Metab* **89**:2078–2084, 2004.

Ahren B. Glucagon-like peptide-1 (GLP-1): a gut hormone of potential interest in the treatment of diabetes. *Bioessays* **20**:642–651, 1998.

Billaudel B, Faure A, Labriji-Mestaghanmi H and Sutter BC. Direct in vitro effect of 1,25-dihydroxyvitamin D3 on islets insulin secretion in vitamin deficient rats: influence of vitamin D3 pretreatment. *Diabetes Metab.* **15**:85–87, 1989.

Bland R, Maskovic D, Hills CE, Hughes SV, Chan SL, Squires PE and Hewison M. Expression of 25-hydroxyvitamin D_3-1-a-hydroxylase in pancreatic islets. *J. Steroid Biochem Mol Biol* **89–90**:121–125, 2004.

Boucher BJ. Inadequate vitamin D status: does it contribute to the disorders comprising syndrome 'X'? *Br J Nutr* **79**:315–327, 1998.

Boucher BJ, Mannan N, Noonan K, Hales CN and Evans SJW. Glucose intolerance and impairment of insulin secretion in relation to vitamin D deficiency in east London Asians. *Diabetologia* **38**:1239–1245, 1995

Bourlon PM, Faure-Dussert A and Billaudel B. The de novo synthesis of numerous proteins is decreased during vitamin D_3 deficiency and is gradually restored by 1,25-dihydroxyvitamin D_3 repletion in the islets of Langerhans of rats. *J Endocrinol* **162**:101–109, 1999.

Burkey BF, Li X, Bolognese L, Balkan B, Mone M, Russell M, Hughes TE and Wang PR. Acute and chronic effects of the incretin enhancer vildagliptin in insulin-resistant rats. *J Pharmacol Exp Ther* **315**:688–695, 2005.

Cade C and Norman AW. Vitamin D3 improves impaired glucose tolerance and insulin secretion in the vitamin D-deficient rat in vivo. *Endocrinology* **119**:84–90, 1986.

Cheng Q, Law PK, De Gasparo M and Leung PS. Combination of DPP-IV inhibitor LAF237 with AT1 receptor antagonist valsartan enhances pancreatic islet morphology and function in a mouse model of type 2 diabetes. *J Pharmacol Exp Ther* **327**:683–691, 2008.

Chiu KC, Chu A, Go VL and Saad MF. Hypovitaminosis D is associated with insulin resistance and beta cell dysfunction. *Am J Clin Nutr* **79**:820–825, 2004.

Chu KY, Lau T, Carlsson PO and Leung PS. Angiotensin II type 1 receptor blockade improves beta-cell function and glucose tolerance in a mouse model of type 2 diabetes. *Diabetes* **55**: 367–374, 2006.

Chu KY and Leung PS. Angiotensin II type 1 receptor antagonism mediates uncoupling protein 2-driven oxidative stress and ameliorates pancreatic islet β-cell function in young type 2 diabetic mice. *Antioxid Redox Signal* **9**:869–878, 2007.

Chu KY, Cheng Q, Chen C, Au LS, Seto SW, Tuo Y, Motin L, Kwan YW and Leung PS. Angiotensin II exerts glucose-dependent effects on K_v currents in mouse pancreatic beta cells via angiotensin II type 2 receptor. *Am J Physiol* **298**:C313–C323, 2010.

Clark SA, Stumpf WE, Sar M and DeLucca HF. 1,25-dihydroxyvitamin D3 target cells in immature pancreatic islets. *Am J Physiol* **253**:E99–E105, 1987.

Connell JM, MacKenzie SM, Freel EM, Fraser R and Davies E. A lifetime of aldosterone excess: long-term consequences of altered regulation of aldosterone production for cardiovascular function. *Endocr Rev* **29**:133–154, 2008.

Dardenne O, Prud'homme J, Arabian A, Glorieux FH and St-Arnaud R. Targeted inactivation of the 25-hydroxyvitamin D(3)-1(alpha)-hydroxylase gene (CYP27B1) creates an animal model of pseudovitamin D-deficiency rickets. *Endocrinology* **142**:3135–3141, 2001.

Davidson HW, Rhodes CJ and Hutton JC. Intra-organellar calcium and pH control proinsulin cleavage in the pancreatic beta cell via two distinct site-specific endopeptidases. *Nature* **333**: 93–96, 1988.

Diz DI, Jessup JA, Westwood BM, Bosch SM, Vinsant S, Gallagher PE, and Averill DB. Angiotensin peptides as neurotransmitters and neuromodulators in the dorsomedial medulla. *Clin Exp Pharmacol Physiol* **29**:473–482, 2002.

Doyle ME and Egan JM. Mechansisms of action of GLP-1 in the pancreas. *Pharmacol Ther* **113**:546–593, 2007.

DREAM Trial Investigators. Effect of ramipril on the incidence of diabetes. *N Engl J Med* **355**:1551–1562, 2006.

Elliott WJ and Meyer PM. Incident diabetes in clinical trials of antihypertensive drugs: a network meta-analysis. *Lancet* **369**:201–207, 2007a (Correction ibid: 1518).

Elliott WJ and Meyer PM. Incident diabetes in clinical trials of antihypertensive drugs: a network meta-analysis. Correction. *Lancet* **369**:1518, 2007b.

Erben RG, Soegiarto DW, Weber K, Zeitz U, Lieberherr M, Gniadecki R, Moller G, Adamski J and Balling R. Deletion of deoxyribonucleic acid binding domain of the vitamin D receptor abrogates genomic and nongenomic functions of vitamin D. *Mol Endocrinol* **16**: 1524–1537, 2002.

Forman JP, Giovannucci E, Holmes MD, Bishcoff-Ferrari HA, Tworoger SS, Willett WC and Curhan GC. Plasma 25-hydroxyvitamin D levels and risk of incident diabetes. *Hypertension* **49**:1063–1069, 2007.

Freundlich M, Quiroz Y, Zhang Z, Zhang Y, Bravo Y, Weisinger JR, Li YC and Rodriguez-Iturbe B. Suppression of renin-angiotensin gene expression in the kidney by paricalcitol. *Kidney Int* **74**:1394–1402, 2008.

Green BD, Liu HK, McCluskey JT, Duffy NA, O'Harte FP, McClenaghan NH, and Flatt, PR. Function of a long-term, GLP-1-treated, insulin-secreting cell line is improved by preventing DPP IV-mediated degradation of GLP-1. *Diabetes Obes Metab* **7**:563–569, 2005.

Hewison M, Burke F, Evans KN, Lammas DA, Sansom DM, Liu P, Modlin RL and Adams JS. Extra-renal 25-hydroxyvitamin D3-1alpha-hydroxylase in human health and disease. *J Steroid Biochem Mol Biol* **103**:316–321, 2007.

Holick MF. Vitamin D: a millenium perspective. *J Cell Biochem* **88**:296–307, 2003.

Holst JJ. Therapy of type 2 diabetes mellitus based on the actions of glucagon-like peptide-1. *Diabetes Metab Res Rev* **18**:430–441, 2002.

Hypponen E, Boucher BJ, Berry DJ and Power C. 25-hydroxyvitamin D, IGF-1, and metabolic syndrome at 45 years of age: a cross-sectional study in the 1958 British Birth Cohort. *Diabetes* **57**:298–305, 2008.

Ishida H and Norman AW. Demonstration of high affinity receptors for 1,25-dihydroxyvitamin D3 in rat pancreas. *Mol Cell Endocrinol* **60**:109–117, 1988.

Kadowski S and Norman AW. Dietary vitamin D is essential for normal insulin secretion from the perfused rat pancreas. *J Clin Invest* 73:759–766, 1984.

Kampf C, Lau T, Olsson R, Leung PS and Carlsson PO. Angiotensin II type 1 receptor inhibition markedly improves the blood perfusion, oxygen tension and first phase of glucose-stimulated insulin secretion in revascularised syngeneic mouse islet grafts. *Diabetologia* **48**: 1159–1167, 2005.

Kieffer TJ, McIntosh CH and Pederson RA. Degradation of glucose-dependent insulinotropic polypeptide and truncated glucagon-like peptide 1 in vitro and in vivo by dipeptidyl peptidase IV. *Endocrinology* **136**:3585–3596, 1995.

Kintscher U, Foryst-Ludwig A and Unger T. Inhibiting angiotensin type 1 receptor as a target for diabetes. *Expert Opin Ther Targets* **12**:1257–1263, 2008.

Ko SH, Kwon HS, Kim SR, Moon SD, Ahn YB, Song KH, Son HS, Cha BY, Lee KW, Son HY, Kang SK, Park CG, Lee IK and Yoon KH. Ramipril treatment suppresses islet fibrosis in Otsuka Long–Evans Tokushima fatty rats. *Biochem Biophys Res Commum* **316**:114–122, 2004.

Ko SH, Hong OK, Kim JW, Ahn YB, Song KH, Cha BY, Son HY, Kim MJ, Jeong IK and Yoon KH. High glucose increases extracellular matrix production in pancreatic stellate cells by activating the renin-angiotensin system. *J Cell Biochem* **98**:343–355, 2006.

Kobayashi K, Sumimoto H and Nawata. Increased expression of NAD(P)H oxidase in islets of animal models of type 2 diabetes and its improvement by an AT1 receptor antagonist. *Biochem Biophys Res Commun* **332**:927–933, 2005.

Kong J and Li YC. Effect of angiotensin II type 1 receptor antagonist and angiotensin-converting inhibitor on vitamin D receptor-null mice. *Am J Physiol* **285**:R255–R261, 2003.

Krause R, Buhring M, Hopfenmuller W, Holick MF and Sharma AM. Ultraviolet B and blood pressure. *Lancet* **352**:709–710, 1998.

Lam SK and Owen A. Incident diabetes in clinical trials of antihypertensive drugs. *Lancet* **369**:1513–1514, 2007.

Lau T, Carlsson PO and Leung PS. Evidence for a local angiotensin-generating system and dose-dependent inhibition of glucose-stimulated insulin release by angiotensin II in isolated pancreatic islets. *Diabetologia* **47**:240–248, 2004.

Lee S, Clark SA, Gill RK and Christakos S. 1,25-dihydroxyvitamin D3 and pancreatic β-cell function: vitamin D receptors, gene expression, and insulin secretion. *Endocrinology* **134**:1602–1610, 1994.

Leung PS and De Gasparo M. Novel peptides and proteins in diabetes mellitus. *Curr Protein Pept Sci* **10**:1–107, 2009.

Leung KK and Leung PS. Effects of hyperglycemia on angiotensin II receptor type expression and insulin secretion in an INS-1E pancreatic beta-cell line. *J Pancreas* **9**:290–299, 2008.

Leung PS. The physiology of a local renin-angiotensin system in the pancreas. *J Physiol* **580**: 31–37, 2007a.

Leung PS. Mechanisms of protective effects induced by blockade of the renin-angiotensin system: novel role of the pancreatic islet angiotensin-generating system in type 2 diabetes. *Diabet Med* **24**:110–116, 2007b.

Leung PS and De Gasparo M. Involvement of the pancreatic renin-angiotensin system in insulin resistance and the metabolic syndrome. *J Cardiometab Syndr* **1**:197–203, 2006.

Leung PS, Hayek A and De Gasparo M. Diabetes: new research and novel therapeutics. *Int J Biochem Cell Biol* **38**:685–1022, 2006.

Li YC, Qiao G, Uskokovic M, Xiang W, Zheng W and Kong J. Vitamin D: a negative endocrine regulator of the renin-angiotensin system and blood pressure. *J Steroid Biochem Mol Biol* **89–90**:387–392, 2004.

Li YC, Kong J, Wei M, Chen ZF, Liu SQ and Cao LP. 1,25-Dihydroxyvitamin D(3) is a negative endocrine regulator of the renin-angiotensin system. *J Clin Invest* **110**:229–238, 2002.

Li YC, Pirro AE, Amling M, Delling G, Baron R, Bronson R and Demay MB. Targeted ablation of the vitamin D receptor: an animal model of vitamin D-dependent rickets type II with alopecia. *Proc Natl Acad Sci USA* **94**:9831–9835, 1997.

Lind L, Wengle B, Wide L and Ljunghall S. Reduction of blood pressure during long-term treatment with active vitamin D (alphacalcidol) is dependent on plasma renin activity and calcium status. A double-blind, placebo-controlled study. *Am. J Hypertens* **2**:20–25, 1989.

Lind L, Lithell H, Skafors E, Wide L and Ljunghall S. Reduction in blood pressure by treatment with alphacidol: a double-blind, placebo-controlled study in subjects with impaired glucose tolerance. *Acta Med Scand* **223**:211–217, 1988.

Lupi R, Guerra SD, Bugliani M, Boggi U, Mosca F, Torri S, Prato SD and Marchetti P. The direct effects of the angiotensin-converting enzyme inhibitors, zofenoprilat and enalaprilat, on isolated human pancreatic islets. *Eur J Endocrinol* **154**:355–361, 2006.

Mathieu C, Gysemans C, Giulietti A and Bouillon R. Vitamin D and diabetes. *Diabetologia* **48**:1247–1257, 2005.

Malloy PJ, Eccleshall TR, Gross C, Van Maldergem L, Bouillon R and Feldman D. Hereditary vitamin D resistant rickets caused by a novel mutation in the vitamin D receptor that results in decreased affinity for hormone and cellular hyporesponsiveness. *J Clin Invest* **99**:297–304, 1997.

Mathieu C, Waer M, Laureys J, Rutgeerts O and Bouillon R. Prevention of autoimmune diabetes in NOD mice by 1,25-dihydroxyvitamin D3. *Diabetologia* **37**:552–558, 1994.

Nakayama M, Inoguchi T, Sonta T, Maeda Y, Sasaki S, Sawada F, Tsubouchi H, Sonoda N, Mest HJ and Mentlein R. Dipeptidyl peptidase inhibitors as new drugs for the treatment of type 2 diabetes. *Diabetologia* **48**:616–620, 2005.

Naslund E, Barkeling B, King N, Gutniak M, Blundell JE, Holst JJ, Rossner S and Hellstrom PM. Energy intake and appetite are suppressed by glucagon-like peptide-1 (GLP-1) in obese men. *Int J Obes Relat Metab Disord* **23**:304–311, 1999.

Norman AW. Vitamin D receptor: new assignments for an already busy receptor. *Endocrinology* **147**:5542–5548, 2006.

Norman AW and Seino Y. An insulinotropic effect of vitamin D analog with increasing intracellular Ca^{2+} concentration in pancreatic β-cells through nongenomic signal transduction. *Endocrinology* **140**:4760–4712, 1999.

ONTARGET Investigators. Telmisartan, ramipril, or both in patients at high risk for vascular events. *N Eng J Med* **358**:1547–1559, 2008.

Paul M, Mehr AP and Kreutz R. Physiology of local renin-angiotensin systems. *Physiol Rev* **86**:747–803, 2006.

PEACE Trial Investigators. The prevention of events with angiotensin converting enzyme inhibition in stable coronary artery disease. *N Eng J Med* **351**:2058–2068, 2004.

Prigeon RL, Quddusi S, Paty B and D'Alessio DA. Suppression of glucose production by GLP-1 independent of islet hormones: a novel extrapancreatic effect. *Am J Physiol Endocrinol Metab* **285**:E701–E707, 2003.

Panda DK, Miao D, Tremblay ML, Sirois J, Farookhi R, Hendy GN and Goltzman D. Targeted ablation of the 25-hydroxyvitamin D 1alpha -hydroxylase enzyme: evidence for skeletal, reproductive, and immune dysfunction. *Proc Natl Acad Sci USA* **98**:7498–7503, 2001.

Repa JJ, Lund EG, Horton JD, Leitersdorf E, Russell DW, Dietschy JM and Turley SD. Disruption of the sterol 27-hydroxylase gene in mice results in hepatomegaly and hypertriglyceridemia. Reversal by cholic acid feeding. *J Biol Chem* **275**:39685–39692, 2000.

Sato H, Uchiyama Y, Masushige S, Fukamizu A, Matsumoto T and Kato S. Mice lacking the vitamin D receptor exhibit impaired bone formation, uterine hypoplasia and growth retardation after weaning. *Nat Genet* **16**:391–396, 1997.

Scheen AJ. Renin-angiotensin system inhibition prevents type 2 diabetes mellitus: a meta-analysis of randomized clinical trials. *Diabetes Metab* **30**:487–496, 2004a.

Scheen AJ. Renin-angiotensin system inhibition prevents type 2 diabetes mellitus: overview of physiological and biochemical mechanisms. *Diabetes Metab* **30**:498–505, 2004b.

Scragg R, Sowers M and Bell C. Serum 25-hydroxyvitamin D, ethnicity, and blood pressure in the Third National Health and Nutrition Examination Survey. *Am J Hypertens* **20**:713–719, 2007.

Scragg R, Sowers M and Bell C. Serum 25-hydroxyvitamin D, diabetes and ethnicity in the Third National Health and Nutrition Examination Survey. *Diabetes Care* **27**:2813–2818, 2004.

Scragg R, Jackson R, Holdaway IM, Lim T and Beaglehole R. Myocardial infarction is inversely associated with plasma 25-hydroxyvitamin D3 levels: a community-based study. *Int J Epidemiol* **19**:559–563, 1990.

Scragg R. Seasonality of cardiovascular disease mortality and the possible protective effect of ultra-violet irradiation. *Int J Epidemiol* **10**:337–341, 1981.

Sergeev IN and Rhoten WB. 1,25-Dihydroxyvitamin D_3 evokes oscillations of intracellular calcium in a pancreatic beta-cell line. *Endocrinology* **136**:2852–2861, 1995.

Shao J, Iwashita N, Ikeda F, Ogihara T, Uchida T, Shimizu T, Uchino H, Hirose T, Kawamori R and Watada H. Beneficial effects of candesartan, an angiotensin II type 1 receptor blocker, on beta-cell function and morphology in db/db mice. *Biochem Biophys Res Commun* **344**:1224–1233, 2006.

Staples JA, Ponsonby AL, Lim L and McMichael AJ. Ecological analysis of some immune-related disorders, including type 1 diabetes, in Australia: latitude, regional ultraviolet radiation, and disease prevalence. *Environ Health Perspect* **111**, 518–523, 2003.

St-Arnaud R, Arabian A, Travers R, Barletta F, Raval-Pandya M, Chapin K, Depovere J, Mathieu C, Christakos S, Demay MB and Glorieux FH. Deficient mineralization of intramembranous bone in vitamin D-24-hydroxylase-ablated mice is due to elevated 1,25-dihydroxyvitamin D and not to the absence of 24,25-dihydroxyvitamin D. *Endocrinology* **141**: 2658–2666, 2000.

Thorens B, Porret A, Buhler L, Deng SP, Morel P and Widmann C. Cloning and function expression of the human islet GLP-1 receptor: demonstration that exendin-4 is an agonist and exendin-(9–39) an antagonist to the receptor. *Diabetes* **42**:1678–1682, 1993.

Thorens B. Expression and cloning of the pancreatic beta cell receptor for the gluco-incretin hormone glucagon-like peptide 1. *Proc Natl Acad Sci USA* **89**:8641–8645, 1992.

Tikellis C, Wookey PJ, Candido R, Andrikopoulos S, Thomas MC and Cooper ME. Improved islet morphology after blockade of the renin-angiotensin system in the ZDF rat. *Diabetes* **53**: 989–997, 2004.

Van Cromphaut SJ, Dewerchin M, Hoenderop JG, Stockmans I, Van Herck E, Kato S, Bindels RJ, Collen D, Carmeliet P, Bouillon R and Carmeliet G. Duodenal calcium absorption in vitamin D receptor-knockout mice: functional and molecular aspects. *Proc Natl Acad Sci USA* **98**: 13324–13329, 2001.

Wang Q and Brubaker PL. Glucagon-like peptide-1 treatment delays the onset of diabetes in 8 week-old db/db mice. *Diabetologia* **45**:1263–1273, 2002.

Wang Q and Jin T. The role of insulin signaling in the development of β-cell dysfunction and diabetes. *Islets* **1**:95–105, 2009.

Weishaar RE and Simpson RU. Vitamin D3 and cardiovascular function in rats. *J Clin Invest* **79**:1706–1712, 1987.

Yoshizawa T, Handa Y, Uematsu Y, Takeda S, Sekine K, Yoshihara Y, Kawakami T, Arioka K, Yusef S, Sleight P, Pogue J et al. Effects of an angiotensin-converting-enzyme inhibitor, ramipril, on cardiovascular events in high risk patients. The Heart Outcomes Prevention Evaluation Study Investigators. *N Eng J Med* **342**:145–153, 2000.

Zander M, Madsbad S, Madsen JL and Holst JJ. Effect of 6-week course of glucagon-like peptide 1 on glycaemic control, insulin sensitivity, and beta-cell function in type 2 diabetes: a parallel-group study. *Lancet* **359**:824–830, 2002.

Chapter 9
Current Research Concerning the RAS in Pancreatic Stem Cells

9.1 Source of Pancreatic Stem Cells

The explosion in the prevalence of diabetes mellitus has expanded beyond the Western world into Asian countries such as China. Given the complication of diabetes, this expanding prevalence carries an attendant medical, social and financial burden for countries with afflicted populations. Central to the pathogenesis of diabetes mellitus is a state of insulin insufficiency, either absolute in the case of type 1 (T1DM) or relative in the case of type 2 (T2DM). The administration of exogenous insulin contributes to the management of T2DM, and has become the mainstay of treatment in T1DM. However, this therapy brings with it a substantial social and lifestyle impact. It is in this context that recent success in the islet transplantation field offers promise for diabetic patients. A lack of availability of human donor islets for transplantation, however, hampers the development and implementation of such novel therapeutic strategies. A new source of such cells must be identified and, in this context, the induction of either embryonic or somatic stem cell differentiation into islet cells offers hope. Embryonic stem cells (ESCs) display features suggesting that their differentiation into pancreatic insulin-producing cells may be possible. However, the true origin of insulin release from these cells is uncertain, and it remains a concern that such cells may have tumorigenic properties. Stem cells derived from the pancreas itself, however, might offer an exciting alternative. That such cells exist suggests that the endocrine pancreas is capable of regeneration long into adulthood. The potential use of pancreatic stem cells (PSCs) as a source of mature islet cells has recently been fuelled by the demonstration that such cells may be induced to differentiate into islet-like cell clusters (ICCs). Nevertheless, these putative islets have so far failed to achieve full maturation.

Tremendous effort has been made in the search for a reliable source of PSCs that would be profoundly expandable in laboratory manipulations and that would have a high differentiation potentiality into functional and transplantable islets. Potential sources for such hormone-producing cells include ESCs, adult pancreas-derived and fetal pancreas-derived stem cells, mesenchymal stem cells (MSCs), and those trans-differentiated from different adult tissues like the liver, spleen and bone marrow (Fig. 9.1). Currently, none of them is considered to be a perfect source for clinical use. Nevertheless, it is appropriate that multiple possibilities be pursued as

P.S. Leung, *The Renin-Angiotensin System: Current Research Progress in The Pancreas*, Advances in Experimental Medicine and Biology 690,
DOI 10.1007/978-90-481-9060-7_9,

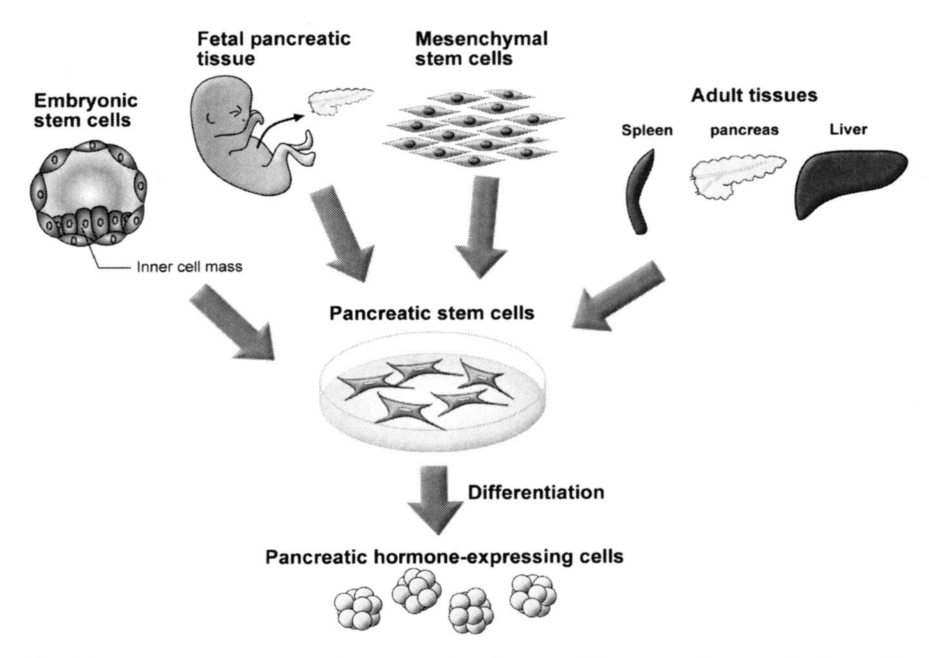

Fig. 9.1 A contemporary approach to generating the potential sources of pancreatic stem cells from embryonic, fetal, mesenchymal and adult tissues that can be proliferated and differentiated into pancreatic insulin-secreting cells

each cell type may have its pros and cons and each will thus contribute to research progress.

Among the various pancreatic progenitors, the potentiality of pluripotent ESCs that are able to differentiate into pancreatic cells is being studied extensively. There have been a number of reports showing differentiation protocols instructing the generation of hormone-producing cells from ESCs (Shim et al., 2004; León-Quinto et al., 2004; D'Amour et al., 2006; Jiang et al., 2007a, b). These protocols are generally designed according to the instructive signals that direct pancreatic organogenesis and β-cell differentiation during human embryonic development. Formation of the definitive endoderm lineage is the critical first step in achieving the derivation of pancreatic epithelium; the ESCs can then be sequentially developed through the stages of foregut endoderm, pancreatic endocrine/exocrine precursors, and finally hormone-secreting endocrine/exocrine cells (Baetge, 2008). The functional performance of the differentiated cells depends on the protocol employed, but some important research breakthroughs rest not only on tuning the accurate signalling factors in order to instruct stable ESC development into the pancreatic lineage, but also on devising better defined conditions in which to maintain cultured ESCs for such studies.

Fetal pancreas represents another promising source of tissue for de novo development of islets. Though progenitors and PSCs are abundant in the fetal pancreas

(Suen et al., 2008; Huang & Tang, 2003), unresolved issues related to accurate isolation of these cells remain. Basic researchers need to find improved markers to more rigorously designate PSC subpopulations within fetal pancreata. Some markers remain controversial, such as nestin, which is an intermediate filament protein that acts as an early marker of neural stem cells (Humphrey et al., 2003, Ueno et al., 2005). The recently identified PSC marker prominin-1 (CD 133) is expressed in hematopoietic as well as endothelial progenitors, while the transcription factor neurogenin 3 (*NGN 3*) is predominantly found in pancreatic β-cell progenitors. Protocols for isolating and characterizating these groups of PSCs from human fetal pancreatic tissues have been reported (Sugiyama et al., 2007a; Oshima et al., 2007; Hori et al., 2008; Koblas et al., 2008). A differentiation cocktail has also been established to direct these PSCs into hormone-producing ICCs (Suen et al., 2008). Studies attempting to characterize the fetal pancreas-derived PSCs have revealed a similar phenotypic expression of markers in MSCs (Gallo et al., 2007). MSCs are multipotent stem cells that bear a high capacity to differentiate into various cell types such as typical adipocytes and oestoblasts (Ruiz & Chen, 2008; Valenti et al., 2008). They are usually isolated and purified by flow cytometry-based cell sorting from tissues such as bone marrow or umbilical cord blood using a number of specific cell surface markers (Sung et al., 2008; Weir et al., 2008). Intriguingly, the adult human pancreas also contains pancreatic MSCs, though their origin is unclear (Davani et al., 2007). There is no known specific marker expressed specifically in the MSC population. Nevertheless, this adherent cell population is amenable to differentiate into pancreatic β-like cells upon exposure to appropriate growth factors. More promisingly, these MSCs, after transplantation into diabetic animals, can adopt a β-cell lineage within the pancreatic microenvironment (Chang et al., 2009).

Generation of β-cells from the trans-differentiation of other adult cells provides an intriguing alternative approach (Tateishi et al., 2008). With suitable genetic engineering, cells can regain differentiation plasticity, a process known as de-differentiation and re-differentiation. This notion has been demonstrated to be feasible in adult liver cells, for example, which undergo a stable transfection with the *pancreatic and duodenal homeobox-1* (*PDX-1*). These PDX-1 gene-transfected cells can follow the pancreatic lineage and develop into insulin-secreting cells in vitro (Zalzman et al., 2005; Ber et al., 2003). Nevertheless, the issue of stem cell plasticity has yet to be investigated. In particular, artifacts and other false positive factors present in cell culture need to be considered. Some resolution of this issue is coming from in vivo experiments showing the existence of in vivo reprogramming of pancreatic exocrine cells into β-cells (Zhou et al., 2008). The trans-differentiation of acinar cells into β-cells has also been reported in culture (Minami & Seino, 2008). The capacity of adult cells to be convertible into pluripotent stem cells, as recently described in induced pluripotent stem cells (iPSs), provides additional support for the possibility of cell reprogramming (Stadtfeld et al., 2008). In summary, Fig. 9.1 provides a schematic diagram of potential sources of pancreatic stem cells derived from various cells and tissues.

9.2 Current Research on Pancreatic Stem Cells

In order to better manipulate the growth and differentiation of PSCs, it will be primarily useful to understand the phenotypic regulation that occurs and the molecular determinants that are present during pancreatic development. Developmental biologists have done some impressive modelling of pancreatic organogenesis (Setty et al., 2008; Bhushan et al., 2001). In the past decade, a growing body of evidence has elaborated upon the transcription dynamics and signalling pathways involved in the development of several pancreatic cell types (Scharfmann et al., 2008; Murtaugh 2007; Ackermann & Gannon, 2007). This valuable information has been garnered mainly from loss-of-function studies in animal models; and the hierarchy of events that occurs in a developing pancreas is being elucidated. Cells in the pancreas and their potential progenitors have been traced for their origins. These experiments have yielded several theories, including the typical epithelial-to-mesenchymal transition (EMT) theory which postulates that an epithelial cell can phenotypically be trans-differentiated into a highly proliferative mesenchymal cell (Moss & Phodes, 2007). Pancreatic β-cells were once believed to undergo reversible EMT and this could be a major mechanism for β-cell regeneration. Although mesenchymal signals were shown to control the development of early pancreatic progenitors (Duvillié et al., 2006; Attali et al., 2007), recent lineage tracing experiments have not demonstrated that such EMT phenomena exist (Morton et al., 2007; Chase et al., 2007; Atouf et al., 2007); instead, it appears that this 'transition' may result from negative selection of epithelial cells by experimental culture conditions in vitro and proliferating pancreatic mesenchymal cells in vivo (Seeberger et al., 2009).

There are many morphogenic or growth factors for PSC differentiation into ICCs with functional implications. Manipulation of PSC development can be a complicated task, and achieving a successful stem-cell based therapy for diabetes rests entirely on devising a complete differentiation protocol for full PSC maturation into physiologically transplantable islets. Morphogenic factors that potentially regulate PSC development have been proposed. One of such factors is the glucagon-like peptide-1 (GLP-1) and its agonist, exendin-4. The role of GLP-1 in regulating adult pancreatic functions has been studied extensively. GLP-1 is an insulinotropic peptide which has potent effects on pancreatic β-cell insulin secretion, proliferation, and cytoprotection (Rickels et al., 2009; Zhang et al., 2007). Moreover, GLP-1/exendin-4 has been reported to be involved in PSC differentiation, maturation and islet transplantation. It was found to promote insulin-secreting ability in ESC-derived insulin-producing cells (Yue et al., 2006; Bai et al., 2005). There has been evidence showing the existence of GLP-1 receptors on fetal pancreas-derived progenitor cells; indeed, GLP-1/exendin-4 has been shown to enhance PSC differentiation into insulin-secreting ICCs (Suen et al., 2008; List & Habener, 2004). Consistent with these findings, an in vivo study of GLP-1 has also shown that it can promote islet graft functionality in fetal ICCs in transplanted diabetic mice (Suen et al., 2006). In that study, blood glucose level, glucose tolerance, and body weight of transplanted diabetic mice treated with exendin-4 improved significantly compared with the transplanted group not subjected to exendin-4 treatment during a

3-month post-transplantation experimental period (Fig. 9.2). In addition, there were revascularization and insulin-producing cells, as evidenced by *Lectins Bandeiraea simplicifolia* and insulin immunoreactivity, respectively, in the graft bearing kidneys of the extendin-4 treatment group (Suen et al., 2006). Thus there is convergent experimental evidence in support of a positive effect of GLP-1/extendin-4 on the growth and survival of immature endocrine precursor cells in foetal mouse ICCs; this beneficial effect is due, at least partly, to enhanced revascularization during islet transplantation.

Another novel class of morphogenic factors that regulate PSC development is the PDZ-domain containing proteins, especially PDZ-domain containing-2 (PDZD2). The role of PDZ-domain-containing proteins in β-cell development has recently received particular attention. Previous studies have shown that a PDZ-domain

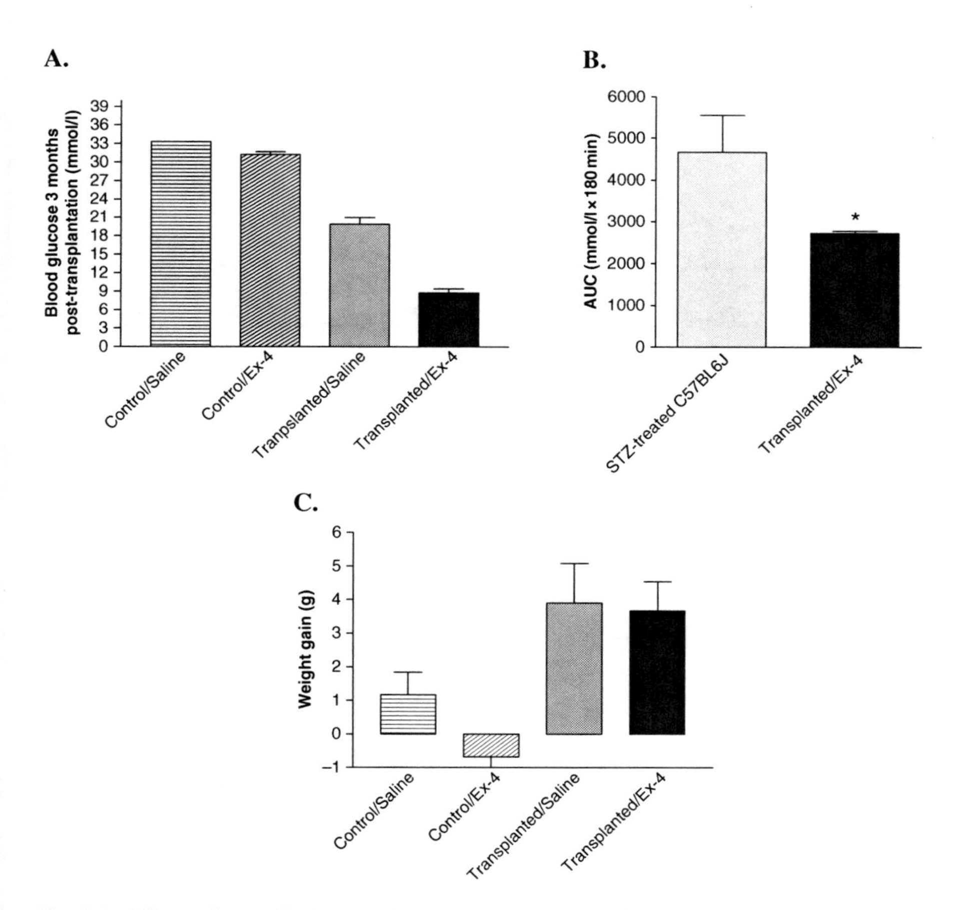

Fig. 9.2 Effects of extendin-4 on **a** blood glucose levels, **b** OGTT and **c** body weight in fetal ICC transplanted mice. *, $p<0.05$, compared with the control group treated with Ex-4. #, $p<0.05$, compared with the transplanted group treated with saline (data are modified from Suen et al., 2008)

coactivator, Bridge-1, interacts directly with the *PDX-1* gene and increases transcriptional activation of the *INSULIN* gene, providing support for the notion that PDZ-domain-containing proteins promote β-cell renewal and restore insulin secretion (Stanojevic et al., 2005). PDZD2 and its secretory form (sPDZD2) were found to be expressed in human foetal pancreata as well as the fetal-derived PSCs (Suen et al., 2008). sPDZD2 expression in PSCs differed according to the gestational week of the fetuses from which the pancreata were derived. When added exogenously to PSC cultures, sPDZD2 exerted a bi-phasic morphogenic effect, concomitant with an inhibitory or stimulatory effect on PSC differentiation at picomolar and nanomolar concentrations, respectively (Fig. 9.3). Moreover, at nanomolar concentrations, the peptide not only augmented the insulin content of the differentiated ICCs, but also triggered the expression of voltage-gated calcium channels that contribute to the insulin-secretory response to membrane depolarization (Leung et al., 2009).

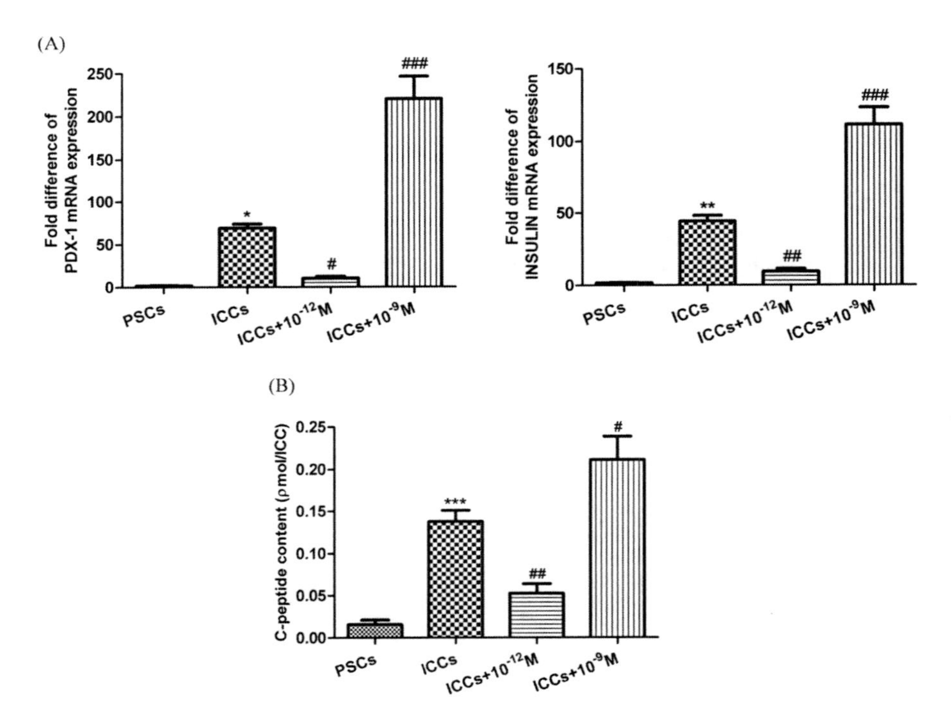

Fig. 9.3 Effect of recombinant sPDZD2 on the expressions of the β-cell phenotypic factors in differentiated ICCs. **a** sPDZD2 caused a concentration-dependent effect on the mRNA expression levels of PDX-1 and Insulin. Expression levels were normalized to β-actin, and relative levels were defined as $2^{-\Delta\Delta CT}$. All data are expressed as the mean ± SEM; n=4 for each group. **b** The C-peptide content of the ICCs was suppressed and enhanced by picomolar and nanomolar concentrations of sPDZD2 treatment respectively. All data are expressed as the mean ± SEM; n=5 for each group. $^{*}p < 0.05$, $^{**}p < 0.01$, $^{***}p < 0.001$ vs. PSCs, $^{\#}p < 0.05$, $^{\#\#}p < 0.01$, $^{\#\#\#}p < 0.001$ vs. control ICCs (data are modified from Leung et al., 2009)

There is thus far no complete protocol for generating from PSCs fully mature ICCs that behave phenotypically and functionally exactly like mature islets. Analysis of the expression of the mature β-cell markers (i.e., *PDX-1*, C-peptide, and the prohormone convertase PC 1/3 that cleaves pro-insulin) is a typical assessment for maturity of differentiated β-cells. There are protocols for generating *PDX-1*- and C-peptide-positive β-cells from ESCs, MSCs and fetal PPCs; however, not all of the products have been assessed for their ability to process proinsulin or their responsivity to glucose challenge (Baetge 2008; Suen et al., 2008; Leung et al., 2009; Davani, et al., 2007). Transplantation of PSC-derived hormone-expressing cells into streptozotocin-induced diabetic animal models has been reported to reverse hyperglycemia in the recipients (Kroon et al., 2008; Wu et al., 2004). Yet very little information has been garnered concerning how long this corrected glycemic condition can be maintained. Scientists are now focusing on how the functionality of these transplanted cells can be promoted as well as on how to maintain their survival. Thorough investigation and understanding in animal studies must be achieved before a large-scale clinical trial on PSCs can be launched. The Food and Drug Administration (FDA) has set out regulations over the administration of any stem cell-based transplantation therapy, with particular focus on the safety and efficacy of the cell products (Halme & Kessler, 2006). A pioneering phase I/II clinical trial was conducted in 2003–2007 in Brazil in which 15 young patients with T1DM were transplanted with stem cells derived from hematopoietic tissues. High-dose immunosupression was applied prior to the transplantation. Insulin independence was achieved in 14 patients for up to 35 months; further studies are in progress with the aim of eliminating the need for anti-rejection drugs (Voltarelli et al., 2007).

The fact that PSCs with differentiation potentiality reside in the adult pancreas has received attention as perhaps pointing to an avenue for β-cell regeneration in vivo (Dor & Melton, 2008). It was once believed that the only mechanism for β-cell turnover was replication of pre-existing β-cells (Teta et al., 2007; Kayali et al., 2007). However, the existence of in vivo trans-differentiation of pancreatic exocrine cells into β-cells was recently substantiated (Zhou et al., 2008). In another experiment using an animal model of injured adult pancreas, resident β-cell progenitors could be activated, thereby leading to an increase in β-cell mass. *NGN 3* may be a useful marker of endogenous progenitors able to give rise to new β-cells (Xu et al., 2008). Such endogenous progenitors may reside in the pancreatic ductal epithelium (Bonner-Weir et al., 2004). These aforementioned experimental data contribute to new concepts for β-cell renewal.

9.3 Current Research of the RAS on Stem Cells

The notion that the RAS is involved in the regulation of embryonic and fetal development of biological organisms is novel. This is distinct from the conventional concept of the RAS acting as a potent vasoconstrictor in the maintenance of blood pressure and fluid homeostasis. Studies designed to examine expression of the RAS components during human embryonic development have been attempted; in this context, major RAS components such as angiotensinogen (Ao), renin and angiotensin-converting enzyme (ACE), as well as type 1 and 2 angiotensin II (AT_1 and AT_2) receptors were found to be expressed in a very early embryonic stage (4th gestational week in humans). The physiologically active peptide angiotensin II was also detected around the same stage, suggesting that it may play a role in organogenesis (Schütz et al., 1996). Recent evidence has also revealed expression of RAS components and their regulation during growth and differentiation of stem cell systems. Accordingly, manipulations of the RAS may alter developmental processes both in vitro and vivo. Table 9.1 tabulates recent studies on RAS involvement in the development of various stem cells, tissues and organs as elaborated below.

9.3.1 The RAS and ESCs

Among the various factors that control the development of ESCs into different cell types, the RAS has emerged as a promising new regulatory candidate. In one study describing the differentiation of human ESCs (H9) along the epithelial lineage, the expression of two receptors, bradykinin B2 and AT_1 receptor, were detected in early stage epithelial cells but not in undifferentiated H9 cells. Addition of angiotensin II and bradykinin led to phosphorylation of extracellular signal-regulated kinase (ERK) 1/2 and Jun N-terminal (JNK) 1/2, thus indicating that the peptides are capable of acting as signalling molecules and therefore perhaps capable of regulating differentiation (Huang et al., 2007). In fact, angiotensin II has been reported to enhance glucose uptake in ESCs (Han et al., 2005) and to exert an AT_1 receptor-mediated mitogenic effect, probably mediated via protein kinase C (PKC) and mitogen-activated protein kinase (MAPK) pathways. Intriguingly, exposure to a high glucose environment in combination with delivery of angiotensin II has been shown to produce a synergistic effect on ESC proliferation in culture (Kim & Han, 2008).

9.3.2 The RAS and MSCs

MSCs have been reported to have the capacity to differentiate into adipocytes (Vidal et al., 2006) and a local RAS has been implicated in the regulation of human MSC differentiation into adipocytes, as first evidenced by the expression of renin, AT_1 receptor, and AT_2 receptor during the developmental process, as well as by a significant increase in the endogenous production of angiotensin II in differentiated cells (Matsushita et al., 2006). Interestingly, endogenous blockade of AT_1 and AT_2 receptor inhibited and promoted adipogenesis of MSCs, respectively, indicating that AT_2 receptor activation may also inhibit MSC differentiation (Matsushita et al., 2006). These data are consistent with clinical observations of RAS blockade acting as a protective agent against the onset of obesity-induced T2DM (Mogi et al., 2006).

Table 9.1 A summary of recent research progress showing RAS involvement in stem cell and tissue/organ development

	Types of cells/tissues	Origins	ANG II-mediated processes	Major RAS components involved					References
				AT_1 receptor	AT_2 receptor	Renin	Ao	ACE	
Stem cells	ESCs	Mouse	Glucose uptake and proliferation	√					Han et al. (2005); Kim & Han (2008)
		Human	Differentiation into epithelial cells	√					Huang et al. (2007)
	MSCs	Human	Differentiation into adipocytes	√	√	√			Matsushita et al. (2006)
	Fetal PSCs	Human	Proliferation	√					Leung et al. (2008)
			Anti-apoptosis		√				Leung et al. (2008)
			Differentiation into hormone-expressing ICCs		√	√	√		Leung et al. (Unpublished data)
Tissues/organs	Vascular system	Mouse; chick	Erythropoiesis	√				√	Kato et al. (2005) and Savary et al. (2005)
		Rat	Senescence in endothelial progenitors	√					Kobayashi et al. (2006)
	Kidney	Mouse; rat	Renal organogenesis	√	√				Sánchez et al. (2008)
	Skin	Human	Epidermal tissue development		√				Liu et al. (2007)
		Human	Wound healing		√				Steckelings et al. (2005)

Other findings related to angiotensin II and MSCs have also shown that this peptide could induce synthesis of vascular endothelial growth factor (VEGF), an angiogenic factor in MSCs (Shi et al., 2009). On the other hand, MSCs have been proposed as a promising alternative to islet transplantation for treating ischemic heart disease and for promoting regeneration of endogenous pancreatic progenitors in patients with diabetes (Dong et al., 2008). Thus, it is possible that angiotensin II-induced production of VEGF might be a contributing underlying mechanism of the beneficial outcomes observed following MSC transplantation (Xu et al., 2008).

9.3.3 The Vascular RAS and Erythropoiesis

In studies using bone marrow-derived endothelial progenitor cells (BM-EPCs) in an angiotensin II-infusion model, angiotensin II has been found to cause senescence in the progenitors by suppressing their differentiation and adherence in culture. This inhibitory effect could be attenuated by administration of an AT_1 receptor antagonist (Kobayashi et al., 2006). In fact, angiotensin II was previously shown to exert an indispensable role during eryothropoiesis (Hubert et al., 2006). In this context, it was found that a significant elevation in levels of plasma erythropoietin (Epo), a glycoprotein hormone that controls erythropoiesis, was observed in transgenic mice expressing human renin and angiotensinogen. Genetic deletion of AT_1 receptor from these mice restored hematocrit levels and reduced Epo levels (Kato et al., 2005). In chick embryos, early expression of ACE was detected in the yolk sac endoderm, which is in contact with aggregates of hemangioblasts in the mesoderm. These findings indicate a potential involvement of angiotensin II in the regulation of blood differentiation. It was shown that ACE inhibition in vivo decreased hematocrit level by 15% (Savary et al., 2005). The notion of ACE and/or angiotensin II being involved in erythropoiesis was further confirmed by a recent study in which ACE marked hematopoietic stem cells from human embryonic, fetal and adult hematopoietic tissues (Jokubaitis et al., 2008).

9.3.4 The RAS and Fetal Tissues

Given the findings of RAS expression in early human embryos and the RAS' potential role in regulation of tissue development, scientists have started to investigate local RAS expression in developing tissues. Human fetal skin, for example, was characterized as having a local RAS. AT_2 receptor was observed at gestational weeks 11–13 and found to be down-regulated with maturation. Conversely, AT_1 receptor was expressed at later developmental stages, with rapidly increasing expression after the 24th week of gestation. Since the AT_2 receptor was expressed mainly in the superficial epidermis, it was believed that it might be involved in fetal skin differentiation processes (Liu et al., 2007). Strong expression of this receptor was also found following injury of the skin in the area of scarring, suggesting that it may be involved in angiotensin II-mediated cutaneous wound healing processes (Steckelings et al., 2005).

Fetal kidney development is also regulated by the RAS. Expression of the *paried homeobox* (*PAX*) genes, *PAX-2* in particular, is critical for normal kidney development. Angiotensin II was once shown to increase the expression of *PAX-2* in fetal renal cells through AT_2 receptor-mediated Janus kinase 2/signal transducers and activators of transcription signalling transduction (JAK2/STAT) pathways (Zhang et al., 2004). Intra-renal AT_2 receptor deficiency can impair PAX-2 gene expression and thus result in congenital anomalies of the kidney and urinary tract (Pope et al., 2001, Chen et al., 2008). AT_1 receptor null mutant mice fail to develop a renal pelvis and ureteral peristaltic movement, a situation which leads to urinary tract obstruction. It is interesting to note that angiotensin II and AT_1 receptor are transiently up-regulated in the vicinity of the renal outlet at birth, perhaps to accommodate the dramatic increase in urine production in the fetus (Pope et al., 2001, Matsusaka et al., 2002). Of note, the AT_1 and AT_2 receptor have recently been shown to counterbalance to each other in the maintenance of normal renal organogenesis (Sánchez et al., 2008). Taken together, these findings indicate that angiotensin II plays a crucial role in normal development of the kidney.

9.4 Current Research on the RAS in Pancreatic Stem Cells

Activation of the local pancreatic RAS has been implicated in (patho)physiology of the pancreas (see Chapter 8 and 10). Yet the notion of a possible link between the RAS and pancreas development, especially with regard to PSCs, is an original and novel avenue in contemporary research. Recent studies using the human fetus-derived PSCs as a model system have strengthened this notion.

A protocol by which one can isolate and characterize a population of PSCs from human fetal pancreata was recently reported by our laboratory. Such PSCs are amenable to differentiate into insulin-secreting ICCs upon stimulation with various morphogenic factors within a period of 7–8 days. This fetus-derived PSC culture could serve as a good platform for future investigation of RAS involvement in regulation of PSC development. In this context, we have shown that the two major receptors of the RAS, namely AT_1 and AT_2 receptor, are expressed on human fetal PSCs, with expression of the AT_1 receptor being particularly strong (Fig. 9.4a). These receptors were found to have differential expression profiles across PSCs derived from different gestational ages (Fig. 9.4b), suggesting that the two receptors may each be involved in pancreatic development at specific time frames (Leung et al., 2008). Addition of angiotensin II exerts a dose-dependent mitogenic effect on the PSCs (Fig. 9.5a), which was found to be mediated by AT_1 receptor (Fig. 9.5b). The precursor of the RAS angiotensinogen and the critical RAS enzyme renin were found to be expressed after initiation of PSC differentiation, but not in undifferentiated PSCs (Leung et al., 2008). The development of a complete local RAS from an incomplete RAS in differentiated ICCs might reflect RAS involvement in

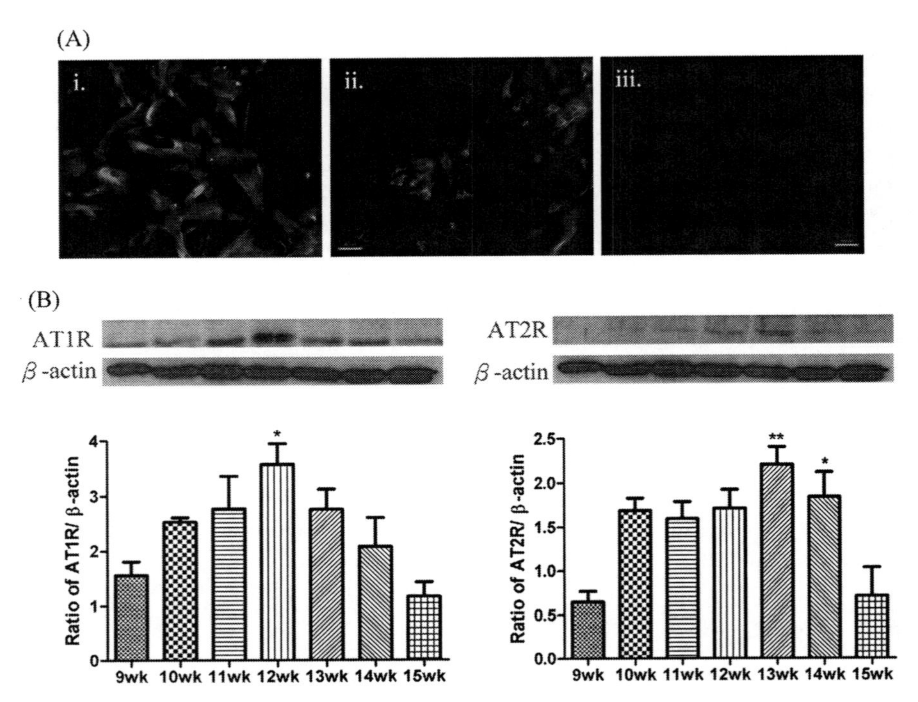

Fig. 9.4 Expression of AT_1 receptor and AT_2 receptor in PSCs. **a** Immunocytochemical detection of (i) AT_1 receptor, (ii) AT_2 receptor on PSCs. The immunoreactivity of AT_1 receptor was more intense than that of AT_2 receptor and (iii) negative control was done by omission of the primary antibodies. Original magnification 200×; Scale bar 100 μM. **b** Protein expression profile of AT_1 receptor and AT_2 receptor in PSCs derived from 9th to 15th gestational week of human fetal pancreata. Data are expressed as the mean ± SEM, n=3 for each group. * $p < 0.05$, ** $p < 0.01$ vs. 9th week PSCs (Data are extracted from Leung et al., 2008)

regulating the differentiation process, as well as being confirmatory of the existence of a functional local RAS in mature adult islets. In fact, angiotensinogen and renin expression peaked early in the PSC differentiation process (i.e. 1st–2nd days), and subsequently exhibited nadirs (Fig. 9.6). Endogenously produced angiotensin II was also detected in the culture medium during differentiation, consistent with involvement in regulation of PSC differentiation (Leung and Leung, unpublished data).

Angiotensin II, the physiologically active peptide of the RAS, has also been studied for its potential effects on the development of fetal PSCs. Our preliminary data suggest that it could exert a dose-dependent mitogenic effect via AT_1 receptor while exhibiting an AT_2 receptor-mediated anti-apoptotic effect on PSCs. In addition, ERK 1/2 and Akt signalling pathways were found to be involved in mediating angiotensin II-induced functions (Leung et al., 2008). Additional preliminary data suggest that angiotensin II may have a stimulatory effect on PSC differentiation into ICCs, as evidenced by the upregulation of *PDX-1* and *INSULIN* mRNA expression (Fig. 9.7). The promotion of PSC differentiation was also found to be mediated, at least in part, by AT_2 receptor, as evidenced by the suppressed C-peptide content of

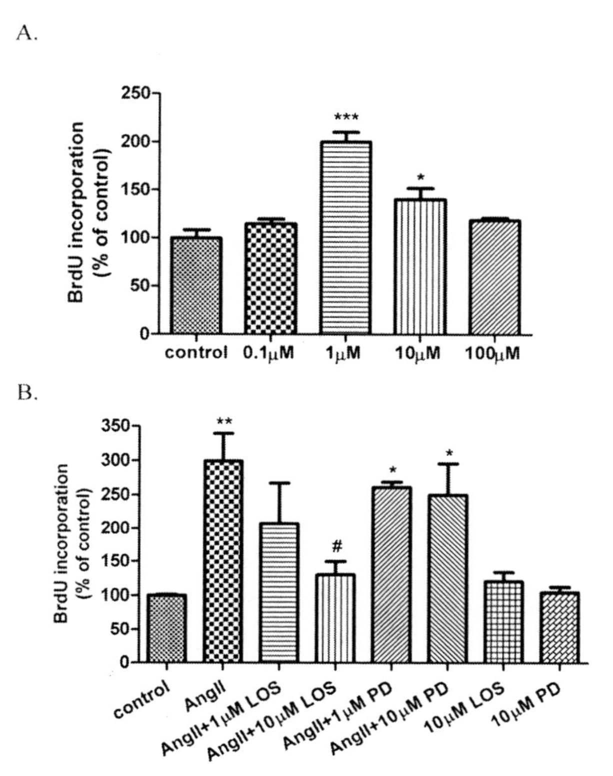

Fig. 9.5 Effect of angiotensin II (Ang II), losartan (Los; AT_1 receptor blocker) and PD 123319 (PD; AT_2 receptor blocker) on the PSC proliferation. Ang II exerted a dose-dependent mitogenic effect on the PSCs, peaking at 1 μM (**a**). This Ang II-induced PSC proliferation could be reversed by pre-treatment with Los but not PD. Addition of Los and PD alone could not exert any observable effects on PSC growth (**b**). All data are expressed as the mean ± SEM; n=3 for each group. $^{*}p$<0.05, ** p<0.01 and $^{***}p < 0.001$ vs. control. $^{\#}p$<0.05 vs. AngII (Data are extracted from Leung et al., 2008)

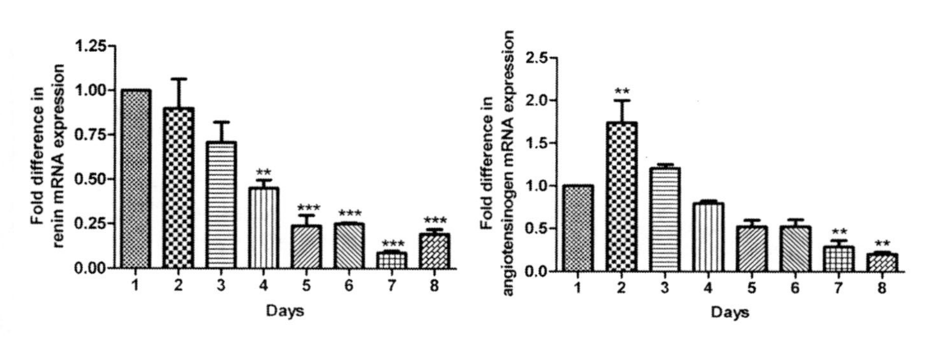

Fig. 9.6 The time-dependent regulated mRNA expression of renin and angiotensinogen during the period of 8-day differentiation of PSCs into ICCs. Expression levels are normalized to β-actin and relative levels are defined as $2^{-\Delta\Delta CT}$. All data are expressed as mean ± SEM; n=3 for each group. Data are expressed as the mean ± SEM, n=5 for each group. $^{*}p < 0.05$, $^{**}p < 0.01$, $^{***}p <$ 0.001 vs. day 1 ICCs (Leung and Leung, unpublished data)

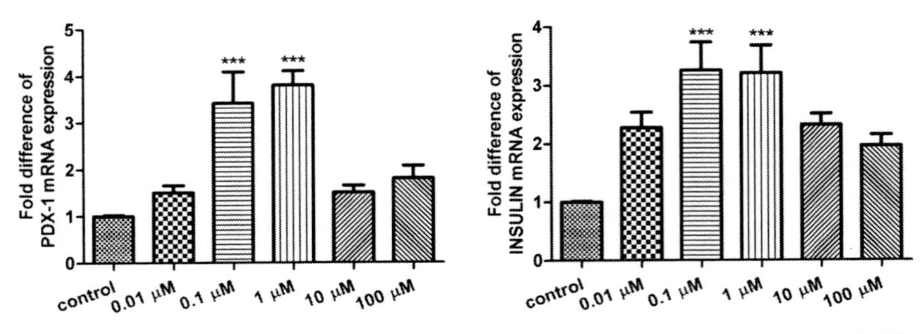

Fig. 9.7 Effect of angiotensin II (Ang II) on the expression of the β-cell phenotypic factors in differentiated ICCs. Ang II exerted a dose-dependent effect on the mRNA expression levels of PDX-1 and INSULIN. Expression levels were normalized to β-actin, and relative levels were defined as $2^{-\Delta\Delta CT}$. All data are expressed as the mean ± SEM; n=3 for each group. ***$p < 0.001$ vs. control ICCs (Leung and Leung, unpublished data)

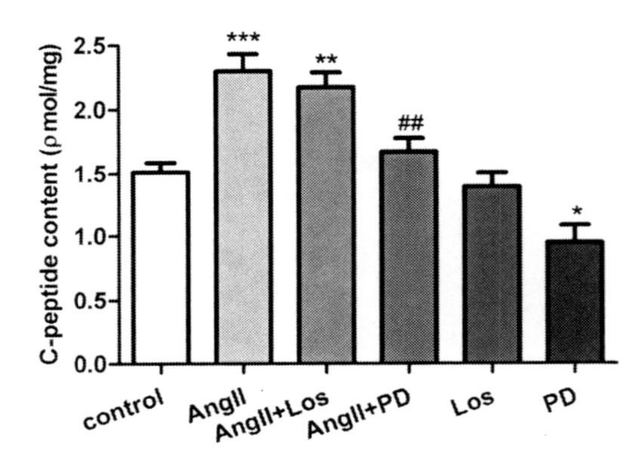

Fig. 9.8 Effect of angiotensin II (Ang II), losartan (Los; AT_1 receptor blocker) and PD 123319 (PD; AT_2 receptor blocker) on the C-peptide content of the differentiated ICCs. Ang II promoted the C-peptide content of the ICCs, and this elevation could be reversed by adding PD but not Los. Addition of PD alone during the PSC differentiation process could suppress the C-peptide content of the ICCs, reflecting the positive role of the endogenous RAS in PSC differentiation. All data are expressed as the mean ± SEM; n=4 for each group. *$p < 0.05$, **$p < 0.01$ and ***$p < 0.001$ vs. control ICCs. ##$p < 0.01$ vs. AngII (Leung and Leung, unpublished data)

the ICCs with the presence of AT_2 receptor blockade during the PSC differentiation process (Fig. 9.8) (Leung and Leung, unpublished data). Further investigation is needed to determine whether these observations occur in vivo to augment the cell mass of residing progenitors and their differentiation. Nevertheless, findings thus far are an important cornerstone with which to reveal an indispensable role of the local RAS in governing pancreatic islet development and regeneration capacity.

To support such a novel idea of linking the local RAS and pancreatic islet development together, our laboratory has attempted to provide preliminary findings by

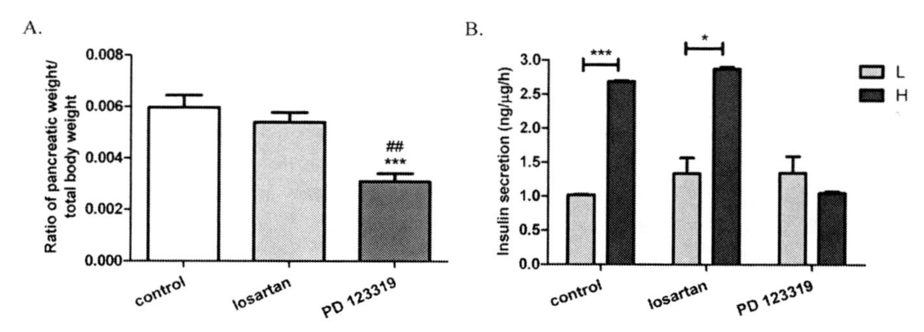

Fig. 9.9 Effects of losartan and PD 123319 application during pregnancy (e8.0-birth) on the development of embryonic pancreas. PD 123319 treatment significantly decreased the weight of neonatal pancreas (**a**) and hampered the glucose-stimulated insulin secretion of its isolated islets (**b**). Data are expressed as the mean ± SEM, $n=4$ in each group. ***$p < 0.001$ vs. control or control (L); ##$p < 0.01$ vs. losartan; *$p < 0.5$ vs. losartan (L) (Leung and Leung, unpublished data)

tracing any abnormal pancreatic development in neonatal animals after manipulating the RAS components during pregnancy. To demonstrate this, we apply losartan and PD 123319, a specific AT_1 and AT_2 receptor blocker, respectively, in drinking water to pregnant mouse mother from e8.0 (where patterning of gut epithelium for future pancreatic buds starts) to birth. Interestingly, we observe a significant decrease in the pancreatic mass in the neonatal mouse pancreas treated with PD 123319 (Fig. 9.9a). Islets isolated from this group of pancreas also give a poorer response to glucose challenge in terms of its lower insulin secretion (Fig. 9.9b). These observations might provide clues for the critical role of AT_2 receptor in regulating normal pancreatic organogenesis.

Given that the RAS appears to be involved in controlling development of various organs and stem cells, manipulation of the RAS offer great potential in the field of stem cell and regenerative medicine. Indeed, there is some evidence that cell development can be controlled through manipulation of the RAS. A rather well-defined area for such a phenomenon is the developmental control of tumour cells, such as prostate and pancreatic cancer cells. RAS interactions with growth factors may mediate cell growth and proliferation (Sanada et al., 2009) and RAS components are upregulated in tumour tissues (Huang et al., 2008). Scientists have tried to designate potential therapeutic implications of the RAS in cancer cells. Observations of AT_1 receptor-mediated angiotensin II induction of tumour growth, angiogenesis and metastasis in experimental models have been reported (Ino et al., 2006; Escobar et al., 2004). Upregulation of AT_2 receptor has been observed in some tumour tissues (Lam & Leung, 2002), and AT_2 receptor antagonism inhibited growth factor signalling pathways in prostate cancer cells (Chow et al., 2008). These data indicate that RAS blockade represents a potential therapeutic approach in oncology. Indeed, some gene polymorphisms observed in certain RAS components, such as angiotensinogen, may confer a higher cancer risk in some populations (Sugimoto et al., 2007b). Interestingly, blockade of the RAS appears to be a promising approach for alleviating several (patho)physiological conditions of the pancreas including diabetes (Chu et al., 2006; Cheng et al., 2008), pancreatitis (Chan & Leung 2007) and pancreatic cancer (Lau & Leung, 2010).

In terms of pancreatic development, β-cell regeneration and PSC differentiation, the RAS may provide a means of elucidating the regulation of these processes. In the fetal pancreas-derived PSC system, development from an incomplete RAS to a complete local RAS after differentiation is particularly noteworthy as it highlights the role of the RAS in such processes (Leung et al., 2008). Existing differentiation protocols for PSCs struggle to generate complete maturation of ICCs that are functionally identical to adult islets. Angiotensin II may be developed as an indispensable morphogen for promoting PSC development. A schematic diagram summarizing the effects of the major RAS components on the regulation of PSC development is shown in Fig. 9.10. The AT_2 receptor has been regarded as a 'pluripotent receptor' that governs the development of many cells and tissues, and hence the angiotensin II-AT_2 receptor axis should be considered a potential pathway for promotion of PSC differentiation. Upon elucidation of the RAS expression

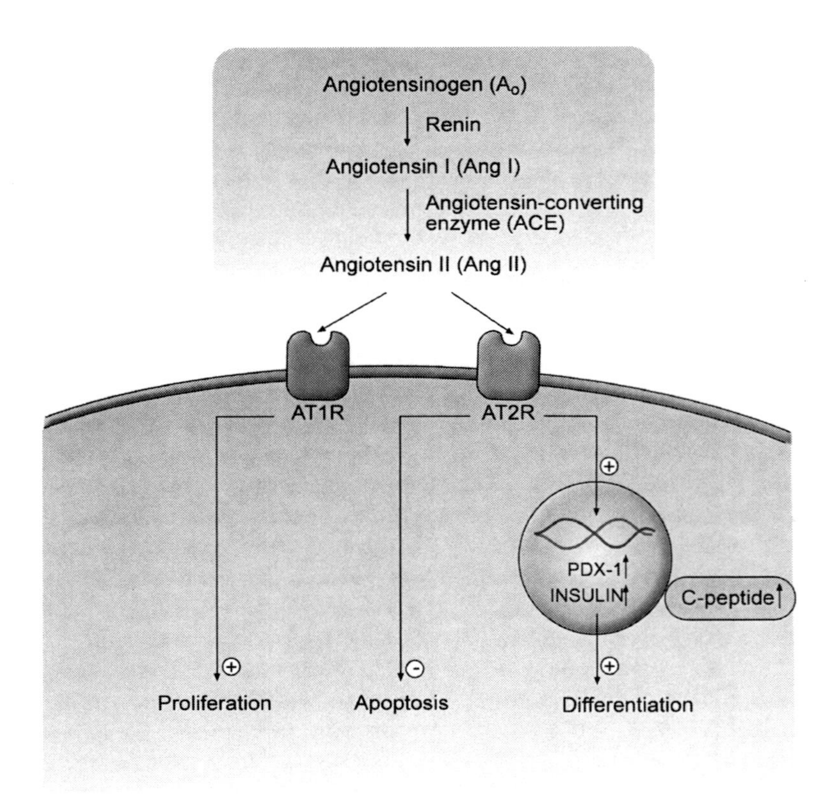

Fig. 9.10 A schematic diagram depicting how angiotensin II (Ang II) mediates its putative effects via AT_1 receptor and AT_2 receptor on human fetal PSC development

dynamics within resident β-cell progenitors, this pathway could also be harnessed to regulate their development and thereby enhance β-cell regeneration in diabetic patients.

Following derivation of functionally active islets from PSCs, the successful implantation and survival of such islets in a host remains an important issue for ongoing and future research. AT_1 receptor blockade during a specific time frame after islet transplantation has been demonstrated to improve islet graft blood perfusion (Kampf et al., 2005). It is noteworthy that RAS blockade has been shown to severely suppress lymphocyte proliferation and cytokine production (Sagawa et al., 2005). This effect may help ameliorate graft rejection by the host immune system. Whether a similar strategy will need to be applied to transplanted PSC-derived islets awaits intensive investigations. Thus far, however, it is clear that the RAS represents a potential means through which to enhance graft functionality and survival.

9.4.1 Limitations and Future Directions

Fully elucidating the complexity of the regulatory network that drives pancreatic development will demand substantial effort and time. The ongoing accrual of new data and ideas will certainly assist in promoting research in the field of PSC biology. A common obstacle in current PSC research is the generation of immature islets in terms of their ability to be highly responsive to glucose challenge and to completely reverse hyperglycemia in diabetic patients after transplantation (Jiang et al., 2007a, b; Phillips et al., 2007). This inherent problem will remain a barrier to performing large-scale clinical trials on human subjects until the nature of PSC-derived islets has been sufficiently characterized.

While the RAS appears to provide a potential list of candidates that may influence PSC research, the underlying mechanism(s) by which pancreatic islet development is controlled remains ambiguous at this stage. Because the RAS is well-defined as a system that regulates various human physiological functions, interfering with it could adversely affect homeostasis. In this context, some clinical studies have reported that RAS blockade in human subjects may cause developmental toxicity during pregnancy as diagnosed by fetal ultrasonography (Serreau et al., 2005). Likewise experiments in animals showed that RAS inhibitors can cross the placental barrier, and thus inhibit fetal RAS (Grove et al., 1995); RAS inhibition can result in organ anomalies in fetuses, such as kidney malformation (Sánchez et al., 2008) and retinopathy (Hård et al., 2008). The fetal effects of RAS blockade during the 2nd and 3rd trimesters of human pregnancy are apparently the greatest, and these fetopathies are probably attributable to a hypoperfusion effect rather than a teratogenic effect (Quan, 2006). Pharmacological blockade of the RAS components might also result in over-activation of other modulatory pathways and thus cause undesirable outcomes (Levy, 2005).

The above discussion covers a number of issues that should be considered when formulating any therapeutic strategies that involve RAS manipulations. An immediate future focus should be placed on the investigation of potential interactions of the RAS with other defined morphogenic factors and/or genetic components in regulating PSC growth and differentiation. Whether therapeutic manipulation of RAS components would support islet regenerative capacity in diabetic patients with upregulated pancreatic RAS has yet to be demonstrated. The data reviewed in this Chapter provide new insights into our understanding of RAS involvement in PSC biology.

References

Ackermann AM and Gannon M. Molecular regulation of pancreatic beta-cell mass development, maintenance, and expansion. *J Mol Endocrinol* **38**:193–206, 2007.

Atouf F, Park CH, Pechhold K, Ta M, Choi Y and Lumelsky NL. No evidence for mouse pancreatic beta-cell epithelial-mesenchymal transition in vitro. *Diabetes* **56**:699–702, 2007.

Attali M, Stetsyuk V, Basmaciogullari A, Aiello V, Zanta-Boussif MA, Duvillie B and Scharfmann R. Control of beta-cell differentiation by the pancreatic mesenchyme. *Diabetes* **56**:1248–1258, 2007.

Baetge EE. Production of beta-cells from human embryonic stem cells. *Diabetes Obes Metab* **10**(Suppl 4):186–194, 2008.

Bai L, Meredith G and Tuch BE. Glucagon-like peptide-1 enhances production of insulin in insulin-producing cells derived from mouse embryonic stem cells. *J Endocrinol* **186**:343–352, 2005.

Ber I, Shternhall K, Perl S, Ohanuna Z, Goldberg I, Barshack I, Benvenisti-Zarum L, Meivar-Levy I and Ferber S. Functional, persistent, and extended liver to pancreas transdifferentiation. *J Biol Chem* **278**:31950–31957, 2003.

Bhushan A, Itoh N, Kato S, Thiery JP, Czernichow P, Bellusci S and Scharfmann R. Fgf10 is essential for maintaining the proliferative capacity of epithelial progenitor cells during early pancreatic organogenesis. *Development* **128**:5109–5117, 2001.

Bonner-Weir S, Toschi E, Inada A, Reitz P, Fonseca SY, Aye T and Sharma A. The pancreatic ductal epithelium serves as a potential pool of progenitor cells. *Pediatr Diabetes* **5**(Suppl):16–22, 2004.

Chan YC and Leung PS. Angiotensin II type 1 receptor-dependent nuclear factor-kappaB activation-mediated proinflammatory actions in a rat model of obstructive acute pancreatitis. *J Pharmacol Exp Ther* **323**:10–18, 2007.

Chang C, Wang X, Niu D, Zhang Z, Zhao H and Gong F. Mesenchymal stem cells adopt beta-cell fate upon diabetic pancreatic microenvironment. *Pancreas* **38**:275–281, 2009.

Chase LG, Ulloa-Montoya F, Kidder BL and Verfaillie CM. Islet-derived fibroblast-like cells are not derived via epithelial-mesenchymal transition from Pdx-1 or insulin-positive cells. *Diabetes* **56**:3–7, 2007.

Chen YW, Tran S, Chenier I, Chan JS, Ingelfinger JR, Inagami T and Zhang SL. Deficiency of intrarenal angiotensin II type 2 receptor impairs paired homeo box-2 and N-myc expression during nephrogenesis. *Pediatr Nephrol* **23**:1769–1777, 2008.

Cheng Q, Law PK, de Gasparo M and Leung PS. Combination of the dipeptidyl peptidase IV inhibitor LAF237 with the angiotensin II type 1 receptor antagonist valsartan enhances pancreatic islet morphology and function in a mouse model of type 2 diabetes. *J Pharmacol Exp Ther* **327**:683–691, 2008.

Chow L, Rezmann L, Imamura K, Wang L, Catt K, Tikellis C, Louis WJ, Frauman AG and Louis SN. Functional angiotensin II type 2 receptors inhibit growth factor signaling in LNCaP and PC3 prostate cancer cell lines. *Prostate* **68**:651–660, 2008.

Chu KY, Lau T, Carlsson PO and Leung PS. Angiotensin II type 1 receptor blockade improves beta-cell function and glucose tolerance in a mouse model of type 2 diabetes. *Diabetes* **55**:367–374, 2006.

D'Amour KA, Bang AG, Eliazer S, Kelly OG, Agulnick AD, Smart NG, Moorman MA, Kroon E, Carpenter MK and Baetge EE. Production of pancreatic hormone-expressing endocrine cells from human embryonic stem cells. *Nat Biotechnol* **24**:1392–1401, 2006.

Davani B, Ikonomou L, Raaka BM, Geras-Raaka E, Morton RA, Marcus-Samuels B and Gershengorn MC. Human islet-derived precursor cells are mesenchymal stromal cells that differentiate and mature to hormone-expressing cells in vivo. *Stem Cells* **25**:3215–3222, 2007.

Dong QY, Chen L, Gao GQ, Wang L, Song J, Chen B, Xu YX and Sun L. Allogeneic diabetic mesenchymal stem cells transplantation in streptozotocin-induced diabetic rat. *Clin Invest Med* **31**:E328–E337, 2008.

Dor Y and Melton DA. Facultative endocrine progenitor cells in the adult pancreas. *Cell* **132**: 183–194, 2008.

Duvillié B, Attali M, Bounacer A, Ravassard P, Basmaciogullari A and Scharfmann R. The mesenchyme controls the timing of pancreatic beta-cell differentiation. *Diabetes* **55**:582–589, 2006.

Escobar E, Rodríguez-Reyna TS, Arrieta O and Sotelo J. Angiotensin II, cell proliferation and angiogenesis regulator: biologic and therapeutic implications in cancer. *Curr Vasc Pharmacol* **2**:385–399, 2004.

Gallo R, Gambelli F, Gava B, Sasdelli F, Tellone V, Masini M, Marchetti P, Dotta F and Sorrentino V. Generation and expansion of multipotent mesenchymal progenitor cells from cultured human pancreatic islets. *Cell Death Differ* **14**:1860–1871, 2007.

Grove KL, Mayo RJ, Forsyth CS, Frank AA and Speth RC. Fosinopril treatment of pregnant rats: developmental toxicity, fetal angiotensin-converting enzyme inhibition, and fetal angiotensin II receptor regulation. *Toxicol Lett* **80**:85–95, 1995.

Halme DG and Kessler DA. FDA regulation of stem-cell-based therapies. *N Engl J Med* **355**: 1730–1735, 2006.

Han HJ, Heo JS and Lee YJ. ANG II increases 2-deoxyglucose uptake in mouse embryonic stem cells. *Life Sci* **77**:1916–1933, 2005.

Hård AL, Wennerholm UB, Niklasson A and Hellström A. Severe ROP in twins after blockage of the renin-angiotensin system during gestation. *Acta Paediatr* **97**:1142–1154, 2008.

Hori Y, Fukumoto M and Kuroda Y. Enrichment of putative pancreatic progenitor cells from mice by sorting for prominin-1 (CD133) and platelet-derived growth factor receptor beta. *Stem Cells* **26**:2912–2920, 2008.

Huang H and Tang X. Phenotypic determination and characterization of nestin-positive precursors derived from human fetal pancreas. *Lab Invest* **83**:539–547, 2003.

Huang W, Yu LF, Zhong J, Qiao MM, Jiang FX, Du F, Tian XL and Wu YL. Angiotensin II type 1 receptor expression in human gastric cancer and induces MMP2 and MMP9 expression in MKN-28 cells. *Dig Dis Sci* **53**:163–168, 2008.

Huang Z, Yu J, Toselli P, Bhawan J, Sudireddy V, Taylor L and Polgar P. Angiotensin II type 1 and bradykinin B2 receptors expressed in early stage epithelial cells derived from human embryonic stem cells. *J Cell Physiol* **211**:816–825, 2007.

Hubert C, Savary K, Gasc JM and Corvol P. The hematopoietic system: a new niche for the renin-angiotensin system. *Nat Clin Pract Cardiovasc Med* **3**:80–85, 2006.

Humphrey RK, Bucay N, Beattie GM, Lopez A, Messam CA, Cirulli V and Hayek A. Characterization and isolation of promoter-defined nestin-positive cells from the human fetal pancreas. *Diabetes* **52**:2519–2525, 2003.

Ino K, Shibata K, Kajiyama H, Nawa A, Nomura S and Kikkawa F. Manipulating the angiotensin system: new approaches to the treatment of solid tumours. *Expert Opin Biol Ther* **6**:243–255, 2006.

Jiang J, Au M, Lu K, Eshpeter A, Korbutt G, Fisk G and Majumdar AS. Generation of insulin-producing islet-like clusters from human embryonic stem cells. *Stem Cells* **25**:1940–1953, 2007a.

Jiang W, Shi Y, Zhao D, Chen S, Yong J, Zhang J, Qing T, Sun X, Zhang P, Ding M, Li D and Deng H. In vitro derivation of functional insulin-producing cells from human embryonic stem cells. *Cell Res* **17**:333–344, 2007b.

Jokubaitis VJ, Sinka L, Driessen R, Whitty G, Haylock DN, Bertoncello I, Smith I, Péault B, Tavian M and Simmons PJ. Angiotensin-converting enzyme (CD143) marks hematopoietic stem cells in human embryonic, fetal, and adult hematopoietic tissues. *Blood* **111**:4055–4063, 2008.

Kampf C, Lau T, Olsson R, Leung PS and Carlsson PO. Angiotensin II type 1 receptor inhibition markedly improves the blood perfusion, oxygen tension and first phase of glucose-stimulated insulin secretion in revascularised syngeneic mouse islet grafts. *Diabetologia* **48**:1159–1167, 2005.

Kato H, Ishida J, Imagawa S, Saito T, Suzuki N, Matsuoka T, Sugaya T, Tanimoto K, Yokoo T, Ohneda O, Sugiyama F, Yagami K, Fujita T, Yamamoto M, Nangaku M and Fukamizu A. Enhanced erythropoiesis mediated by activation of the renin-angiotensin system via angiotensin II type 1a receptor. *FASEB J* **19**:2023–2035, 2005.

Kayali AG, Flores LE, Lopez AD, Kutlu B, Baetge E, Kitamura R, Hao E, Beattie GM and Hayek A. Limited capacity of human adult islets expanded in vitro to redifferentiate into insulin-producing beta-cells. *Diabetes* **56**:703–718, 2007.

Kim YH and Han HJ. Synergistic effect of high glucose and ANG II on proliferation of mouse embryonic stem cells: involvement of PKC and MAPKs as well as AT1 receptor. *J Cell Physiol* **215**:374–382, 2008.

Kobayashi K, Imanishi T and Akasaka T. Endothelial progenitor cell differentiation and senescence in an angiotensin II-infusion rat model. *Hypertens Res* **29**:449–355, 2006.

Koblas T, Pektorova L, Zacharovova K, Berkova Z, Girman P, Dovolilova E, Karasova L and Saudek F. Differentiation of CD133-positive pancreatic cells into insulin-producing islet-like cell clusters. *Transplant Proc* **40**:415–428, 2008.

Kroon E, Martinson LA, Kadoya K, Bang AG, Kelly OG, Eliazer S, Young H, Richardson M, Smart NG, Cunningham J, Agulnick AD, D'Amour KA, Carpenter MK and Baetge EE. Pancreatic endoderm derived from human embryonic stem cells generates glucose-responsive insulin-secreting cells in vivo. *Nat Biotechnol* **26**:443–452, 2008.

Lam KY and Leung PS. Regulation and expression of a renin-angiotensin system in human pancreas and pancreatic endocrine tumours. *Eur J Endocrinol* **146**:567–572, 2002.

Lau ST and Leung PS. Role of the RAS in pancreatic cancer. *Curr Cancer Drug Targets* (in press), 2010.

León-Quinto T, Jones J, Skoudy A, Burcin M and Soria B. In vitro directed differentiation of mouse embryonic stem cells into insulin-producing cells. *Diabetologia* **47**:1442–1451, 2004.

Leung KK, Ma MT and Leung PS. Involvement of the renin-angiotensin system in the growth and differentiation of human pancreatic progenitor cells. *Gordon Research Conferences – Angiotensin*. 24–29 February, 2008. Ventura, CA, 2008.

Leung KK, Suen PM, Lau TK, Ko WH, Yao KM and Leung PS. PDZ-domain containing-2 (PDZD2) drives the maturity of human fetal pancreatic progenitor-derived islet-like cell clusters with functional responsiveness against membrane depolarization. *Stem Cells Dev* **18**:979–990, 2009.

Levy BI. How to explain the differences between renin angiotensin system modulators. *Am J Hypertens* **18**:134S–141S, 2005.

List JF and Habener JF. Glucagon-like peptide 1 agonists and the development and growth of pancreatic beta-cells. *Am J Physiol* **286**:E875–E891, 2004.

Liu HW, Cheng B, Fu XB, Sun TZ and Li JF. Characterization of AT1 and AT2 receptor expression profiles in human skin during fetal life. *J Dermatol Sci* **46**:221–235, 2007.

Matsusaka T, Miyazaki Y and Ichikawa I. The renin angiotensin system and kidney development. *Annu Rev Physiol* **64**:551–561, 2002.

Matsushita K, Wu Y, Okamoto Y, Pratt RE and Dzau VJ. Local renin angiotensin expression regulates human mesenchymal stem cell differentiation to adipocytes. *Hypertension* **48**:1095–1102, 2006.

Minami K and Seino S. Pancreatic acinar-to-beta cell transdifferentiation in vitro. *Front Biosci* **13**:5824–5837, 2008.

Mogi M, Iwai M and Horiuchi M. Emerging concept of adipogenesis regulation by the renin-angiotensin system. *Hypertension* **48**:1020–1032, 2006.

Morton RA, Geras-Raaka E, Wilson LM, Raaka BM and Gershengorn MC. Endocrine precursor cells from mouse islets are not generated by epithelial-to-mesenchymal transition of mature beta cells. *Mol Cell Endocrinol* **270**:87–93, 2007.

Moss LG and Rhodes CJ. Beta-cell regeneration: epithelial mesenchymal transition pre-EMTpted by lineage tracing? *Diabetes* **56**:281–292, 2007.

Murtaugh LC. Pancreas and beta-cell development: from the actual to the possible. *Development* **134**:427–438, 2007.

Oshima Y, Suzuki A, Kawashimo K, Ishikawa M, Ohkohchi N and Taniguchi H. Isolation of mouse pancreatic ductal progenitor cells expressing CD133 and c-Met by flow cytometric cell sorting. *Gastroenterology* **132**:720–732, 2007.

Phillips BW, Hentze H, Rust WL, Chen QP, Chipperfield H, Tan EK, Abraham S, Sadasivam A, Soong PL, Wang ST, Lim R, Sun W, Colman A and Dunn NR. Directed differentiation of human embryonic stem cells into the pancreatic endocrine lineage. *Stem Cells Dev* **16**:561–578, 2007.

Pope JC, Brock JW, Adams MC, Miyazaki Y, Stephens FD and Ichikawa I. Congenital anomalies of the kidney and urinary tract: role of the loss of function mutation in the pluripotent angiotensin type 2 receptor gene. *J Urol* **165**:196–202, 2001.

Pope JC, Nishimura H and Ichikawa I. Role of angiotensin in the development of the kidney and urinary tract. *Nephrologie* **19**:433–446, 1998.

Quan A. Fetopathy associated with exposure to angiotensin converting enzyme inhibitors and angiotensin receptor antagonists. *Early Hum Dev* **82**:233–238, 2006.

Rickels MR, Mueller R, Markmann JF and Naji A. Effect of glucagon-like peptide-1 on β- and α-cell function in isolated islet and whole pancreas transplant recipients. *J Clin Endocrinol Metab* **94**:181–189, 2009.

Ruiz SA and Chen CS. Emergence of patterned stem cell differentiation within multicellular structures. *Stem Cells* **26**:2921–2927, 2008.

Sagawa K, Nagatani K, Komagata Y and Yamamoto K. Angiotensin receptor blockers suppress antigen-specific T cell responses and ameliorate collagen-induced arthritis in mice. *Arthritis Rheum* **52**:1920–1928, 2005.

Sanada F, Taniyama Y, Iekushi K, Azuma J, Okayama K, Kusunoki H, Koibuchi N, Doi T, Aizawa Y and Morishita R. Negative action of hepatocyte growth factor/c-Met system on angiotensin II signaling via ligand-dependent epithelial growth factor receptor degradation mechanism in vascular smooth muscle cells. *Circ Res* **105**:667–675, 2009.

Sánchez SI, Seltzer AM, Fuentes LB, Forneris ML and Ciuffo GM. Inhibition of angiotensin II receptors during pregnancy induces malformations in developing rat kidney. *Eur J Pharmacol* **588**:114–123, 2008.

Savary K, Michaud A, Favier J, Larger E, Corvol P and Gasc JM. Role of the renin-angiotensin system in primitive erythropoiesis in the chick embryo. *Blood* **105**:103–110, 2005.

Scharfmann R, Duvillie B, Stetsyuk V, Attali M, Filhoulaud G and Guillemain G. Beta-cell development: the role of intercellular signals. *Diabetes Obes Metab* **10**(Suppl 4):195–200, 2008.

Schütz S, Le Moullec JM, Corvol P and Gasc JM. Early expression of all the components of the renin-angiotensin-system in human development. *Am J Pathol* **149**:2067–2079, 1996.

Seeberger KL, Eshpeter A, Rajotte RV and Korbutt GS. Epithelial cells within the human pancreas do not coexpress mesenchymal antigens: epithelial-mesenchymal transition is an artifact of cell culture. *Lab Invest* **89**:110–121, 2009.

Serreau R, Luton D, Macher MA, Delezoide AL, Garel C and Jacqz-Aigrain E. Developmental toxicity of the angiotensin II type 1 receptor antagonists during human pregnancy: a report of 10 cases. *BJOG* **112**:710–722, 2005.

Setty Y, Cohen IR, Dor Y and Harel D. Four-dimensional realistic modeling of pancreatic organogenesis. *Proc Natl Acad Sci USA* **105**:20374–20389, 2008.

Shi RZ, Wang JC, Huang SH, Wang XJ and Li QP. Angiotensin II induces vascular endothelial growth factor synthesis in mesenchymal stem cells. *Exp Cell Res* **315**:10–25, 2009.

Shim JH, Kim SE, Woo DH, Kim SK, Oh CH, McKay R and Kim JH. Directed differentiation of human embryonic stem cells towards a pancreatic cell fate. *Diabetologia* **50**:1228–1238, 2004.

Stadtfeld M, Brennand K and Hochedlinger K. Reprogramming of pancreatic beta cells into induced pluripotent stem cells. *Curr Biol* **18**:890–904, 2008.

Stanojevic V, Yao KM and Thomas MK. The coactivator Bridge-1 increases transcriptional activation by pancreas duodenum homeobox-1 (PDX-1). *Mol Cell Endocrinol* **237**:67–74, 2005.

Steckelings UM, Henz BM, Wiehstutz S, Unger T and Artuc M. Differential expression of angiotensin receptors in human cutaneous wound healing. *Br J Dermatol* **153**:887–893, 2005.

Suen PM, Li K, Chan JC and Leung PS. In vivo treatment with glucagon-like peptide 1 promotes the graft function of fetal islet-like cell clusters in transplanted mice. *Int J Biochem Cell Biol* **38**:951–960, 2006.

Suen PM, Zou C, Zhang YA, Lau TK, Chan J, Yao KM and Leung PS. PDZ-domain containing-2 (PDZD2) is a novel factor that affects the growth and differentiation of human fetal pancreatic progenitor cells. *Int J Biochem Cell Biol* **40**:789–803, 2008.

Sugiyama T, Rodriguez RT, McLean GW and Kim SK. Conserved markers of fetal pancreatic epithelium permit prospective isolation of islet progenitor cells by FACS. *Proc Natl Acad Sci USA* **104**:175–180, 2007a.

Sugimoto M, Furuta T, Shirai N, Kodaira C, Nishino M, Ikuma M, Sugimura H and Hishida A. Role of angiotensinogen gene polymorphism on Helicobacter pylori infection-related gastric cancer risk in Japanese. *Carcinogenesis* **28**:2036–2050, 2007b.

Sung JH, Yang HM, Park JB, Choi GS, Joh JW, Kwon CH, Chun JM, Lee SK and Kim SJ. Isolation and characterization of mouse mesenchymal stem cells. *Transplant Proc* **40**:2629–2654, 2008.

Tateishi K, He J, Taranova O, Liang G, D'Alessio AC and Zhang Y. Generation of insulin-secreting islet-like clusters from human skin fibroblasts. *J Biol Chem* **283**:31601–31607, 2008.

Teta M, Rankin MM, Long SY, Stein GM and Kushner JA. Growth and regeneration of adult beta cells does not involve specialized progenitors. *Dev Cell* **12**:817–826, 2007.

Ueno H, Yamada Y, Watanabe R, Mukai E, Hosokawa M, Takahashi A, Hamasaki A, Fujiwara H, Toyokuni S, Yamaguchi M, Takeda J and Seino Y. Nestin-positive cells in adult pancreas express amylase and endocrine precursor cells. *Pancreas* **31**:126–131, 2005.

Valenti MT, Dalle Carbonare L, Donatelli L, Bertoldo F, Zanatta M and Lo Cascio V. Gene expression analysis in osteoblastic differentiation from peripheral blood mesenchymal stem cells. *Bone* **43**:1084–1092, 2008.

Vidal MA, Kilroy GE, Johnson JR, Lopez MJ, Moore RM and Gimble JM. Cell growth characteristics and differentiation frequency of adherent equine bone marrow-derived mesenchymal stromal cells: adipogenic and osteogenic capacity. *Vet Surg* **35**:601–610, 2006.

Voltarelli JC, Couri CE, Stracieri AB, Oliveira MC, Moraes DA, Pieroni F, Coutinho M, Malmegrim KC, Foss-Freitas MC, Simões BP, Foss MC, Squiers E and Burt RK. Autologous nonmyeloablative hematopoietic stem cell transplantation in newly diagnosed type 1 diabetes mellitus. *JAMA* **297**:1568–1576, 2007.

Weir C, Morel-Kopp MC, Gill A, Tinworth K, Ladd L, Hunyor SN and Ward C. Mesenchymal stem cells: isolation, characterisation and in vivo fluorescent dye tracking. *Heart Lung Circ* **17**:395–403, 2008.

Wu F, Jagir M and Powell JS. Long-term correction of hyperglycemia in diabetic mice after implantation of cultured human cells derived from fetal pancreas. *Pancreas* **29**:e23–e39, 2004.

Xu X, D'Hoker J, Stangé G, Bonné S, De Leu N, Xiao X, Van de Casteele M, Mellitzer G, Ling Z, Pipeleers D, Bouwens L, Scharfmann R, Gradwohl G and Heimberg H. Beta cells can be generated from endogenous progenitors in injured adult mouse pancreas. *Cell* **132**:197–207, 2008.

Xu YX, Chen L, Wang R, Hou WK, Lin P, Sun L, Sun Y and Dong QY. Mesenchymal stem cell therapy for diabetes through paracrine mechanisms. *Med Hypotheses* **71**:390–403, 2008.

Yue F, Cui L, Johkura K, Ogiwara N and Sasaki K. Glucagon-like peptide-1 differentiation of primate embryonic stem cells into insulin-producing cells. *Tissue Eng* **12**:2105–2116, 2006.

Zalzman M, Anker-Kitai L and Efrat S. Differentiation of human liver-derived, insulin-producing cells toward the beta-cell phenotype. *Diabetes* **54**:2568–2575, 2005.

Zhang J, Tokui Y, Yamagata K, Kozawa J, Sayama K, Iwahashi H, Okita K, Miuchi M, Konya H, Hamaguchi T, Namba M, Shimomura I and Miyagawa JI. Continuous stimulation of human glucagons-like peptide-1 (7–36) amide in a mouse model (NOD) delays onset of autoimmune type 1 diabetes. *Diabetologia* **50**:1900–1909, 2007.
Zhang SL, Moini B and Ingelfinger JR. Angiotensin II increases Pax-2 expression in fetal kidney cells via the AT2 receptor. *J Am Soc Nephrol* **15**:1452–1465, 2004.
Zhou Q, Brown J, Kanarek A, Rajagopal J and Melton DA. In vivo reprogramming of adult pancreatic exocrine cells to beta-cells. *Nature* **455**:627–632, 2008.

Chapter 10
Current Research of the RAS in Pancreatitis and Pancreatic Cancer

10.1 Basic Studies of the RAS in Pancreatitis

A local RAS exists in several cell types of the pancreas, which exhibits both exocrine and endocrine activities, and this functional local RAS is responsive to various physiological and pathophysiological stimuli (see Chapter 6). Of particular interest in this context are the expression and localization of key RAS components in the acinar and endothelial cells of the exocrine pancreas; these RAS components are subject to upregulation in response to chronic hypoxia and acute pancreatitis (AP). In this regard, an enhanced sensitivity of angiotensin II-mediated vasoconstriction in pancreatic microcirculation could trigger severe ischemia and/or hypoxia conditions, which result in oxidative stress-induced expression of proinflammatory factors, ultimately leading to pancreatic cell inflammation and injury. Indeed, recent investigations have demonstrated that inhibition of RAS activation supports amelioration of pancreatic oxidative stress and tissue injury in experimentally induced AP.

The regulatory mechanism(s) whereby RAS blockade improves pancreatic inflammation and injury, however, have yet to be empirically resolved. Accordingly, the first part of this Chapter provides a critical review of current progress in research concerning the RAS and pancreatic inflammation, with particular focus on the functional role of the RAS-mediated processes in AP. The regulatory pathways involved in RAS inhibition-induced protective effects on pancreatitis-mediated local blood flow, oxidative stress and cytokine-mediated proinflammatory reactions in the pancreas will also be discussed. Data arising from these studies should provide mechanism-driven insight into the novel role for pancreatic RAS antagonism in the prevention and management of AP and its associated inflammation conditions, such as systemic inflammatory response syndrome (SIRS) and multiple organ dysfunction syndrome (MODS). To this end, selective RAS blockers have the potential of providing a novel strategy for future treatment and management of pancreatic inflammation and its systemic complications commonly observed at the tissue and organ levels.

P.S. Leung, *The Renin-Angiotensin System: Current Research Progress in The Pancreas*, Advances in Experimental Medicine and Biology 690,
DOI 10.1007/978-90-481-9060-7_10,

10.1.1 Evidence of a Local RAS and Its Roles in Pancreatitis

The notion of a local RAS in the pancreas was previously proposed in the dog (Chappell et al., 1991, 1992) and in rodents (Ghiani & Masini, 1995; Leung et al., 1997, 1998, 1999), as well as in humans (Tahmasebi et al., 1999; Lam & Leung, 2002). The expression and regulation of a local pancreatic RAS and its potential roles in inflammation have been critically reviewed (Leung & Carlsson, 2001; Leung & Chappell, 2003; Leung, 2003, 2005, 2007). In the exocrine pancreas, angiotensin II receptors are predominantly localized in the pancreatic ducts, blood vessels, and acinar cells (Leung et al., 1997, 1999). Our recent studies showed that experimental AP upregulates RAS components in the pancreas (Leung et al., 2000; Ip et al., 2003a). Prophylactic administration of saralasin, a nonspecific antagonist for type 1 and 2 angiotensin (AT_1 receptor and AT_2 receptor) receptors is protective against AP (Tsang et al., 2003). Activation of pancreatic RAS may augment oxidative stress, leading to pancreatic injury (Ip et al., 2003b; Tsang et al., 2004a). Meanwhile, the AT_1 receptor antagonist losartan can ameliorate cerulein-induced pancreatic and systemic inflammation, including pulmonary inflammation (Chan & Leung, 2006). RAS inhibition with an angiotensin-converting enzyme (ACE) blocker also improves inflammation in chronic pancreatitis, further implicating pancreatic RAS in pancreatic inflammation and fibrosis (Kuno et al., 2003). Although RAS has been shown to be a key mediator in the molecular regulation of cell and tissue inflammation (see review by Suzuki et al., 2003), the mechanistic pathways by which RAS blockade yields protective effects and RAS mediates pancreatic inflammation and injury remain elusive.

10.1.2 Potential Mechanisms of RAS-Mediated Oxidative Stress and Blood Flow in Pancreatitis

Emerging data lends support to the involvement of reactive oxygen species (ROS) in the pathogenesis of the onset of AP (Czako et al., 2000; Rau et al., 2001; Telek et al., 2001; see also recent review by Leung & Chan, 2009). The ROS source in AP may be polymorphonuclear neutrophils, macrophages, and/or endothelial cells through activation of the xanthine–xanthine oxidase system (Schulz et al., 1999; Granell et al., 2003). Hence, there may be activation of a pancreatic RAS in the exocrine pancreas generating ROS during an episode of AP attack. Indeed, angiotensin II stimulates production of superoxide and hydrogen peroxide via the NADPH oxidase system, a major source of ROS in various cells and tissues (Jaimes et al., 1998; Dijkhorst-Oei et al., 1999). NADPH oxidase is predominantly located in neutrophils, and is also in non-phagocytic cells including vascular endothelial cells; it can be stimulated by angiotensin II and cytokines (Griendling et al., 2000). The NADPH oxidase system consists of membrane-bound subunits, namely $gp91^{phox}$ and $p22^{phox}$, and several cytosolic subunits, namely $p40^{phox}$, $p47^{phox}$, $p67^{phox}$ and Rac (Babior, 1999). Stimulation by angiotensin II leads to activation of NADPH oxidase, resulting in the generation of superoxide (Bendall et al., 2002; Dang et al., 2003; Li & Shah, 2003). There may be a close association between RAS activation and NADPH oxidase-dependent generation of ROS in the pancreas. Interestingly, we have demonstrated that activation of the pancreatic RAS by AP is linked to stimulation of NADPH oxidase activity in the pancreas (Tsang et al., 2004a). The regulatory mechanism by which the NADPH oxidase is regulated by the pancreatic RAS during episodes of AP, however, is still largely undefined.

Changes in pancreatic microcirculation, such as vasoconstriction, capillary stasis, decreased oxygen tension, and progressive ischemia are closely associated with AP (Knoefel et al., 1994). As the RAS is a key mediator of potent vasoconstriction and tissue inflammatory response (see reviews by De Gasparo et al., 2000; Suzuki et al., 2003), its selective upregulation by hypoxia and AP should have clinical relevance, both in normal physiology and in pathophysiology, wherein cell and tissue injury in the pancreas result (Chan et al., 2000; Leung et al., 2000; Ip et al., 2002, 2003a).

There is a close interaction between nitric oxide (NO) and RAS in the modulation of blood flow and oxidative stress (De Gasparo, 2002), particularly via the opposing actions of the AT_1 and AT_2 receptors (De Gasparo & Siragy, 1999). Superoxide produced from activated NADPH oxidase is a major cause of NO degradation, and superoxide reacts with NO to form peroxynitrite, a potent nitrosating agent (Beckman & Koppenol, 1996). Peroxynitrite reacts with residual NO, further depleting NO (Kelm et al., 1997). Furthermore, superoxide is readily metabolized into hydrogen peroxide, resulting in NO synthase uncoupling and production of extra superoxide (De Gasparo, 2002). These reactions reduce the bioavailability of NO and enhance the vasoconstrictor effect of angiotensin II, leading to tissue ischemia/hypoxia. The intricate interactions of angiotensin II and NO in regulating pancreatic blood flow and ROS formation, and thus the clinical relevance to AP, remains equivocal.

10.1.3 The RAS and Cytokine-Mediated Systemic Inflammation in Pancreatitis

Systemic inflammation during pancreatitis involves the production of an array of pro-inflammatory and inflammatory mediators (Heath et al., 1993; Grewal et al., 1994). The RAS has been shown to participate in the induction of various pro-inflammatory mediators, such as adhesion molecules, chemokines, cytokines and transcription factors (see review by Suzuki et al., 2003). A local increase in vascular permeability is an early event in eliciting inflammatory reactions and angiotensin II stimulates the synthesis and secretion of vascular permeability factor (Williams et al., 1995; Pupilli et al., 1999). Increased vascular permeability permits cell infiltration and exudation of plasma macromolecules. Recruitment of inflammatory cells to the injury site by inflammatory cytokines is a critical step in the progression of cell inflammation. In this context, angiotensin II-induced ROS have been shown to stimulate the synthesis of adhesion molecules and chemokines (Schieffer et al., 2000), as well as to enhance the migration and adhesion of monocytes and leukocytes (Kintscher et al., 2001; Alvarez & Sanz, 2001). Pancreatitis triggers the expression of tumour necrosis factor-α (TNF-α) in Kupffer cells of the liver via the activation of nuclear factor-kappa B (NF-κB) (Murr et al., 2002). Upon its release into the bloodstream, TNF-α activates the expression of P-selectin (an adhesion molecule) and recruitment of neutrophils in the lungs, eventually resulting in lung damage (Folch-Puy et al., 2003). Coincidentally, it is well documented that angiotensin II has a direct stimulatory effect on the production of NF-κB and TNF-α (Ferreri et al., 1998; Kalra et al., 2002). And interestingly, a functional RAS is necessary for TNF-α-induced cell injury in pulmonary alveolar cells (Wang et al., 2000). Taken together, these findings suggest that that pharmacological blockade of an activated RAS could be a potential therapy for the treatment and management of pancreatitis (Leung, 2003). Nevertheless, the regulatory mechanism(s) by which RAS blockade is protective against pancreatitis-induced SIRS and MODS, however, remain to be elucidated.

10.2 Current Research of the RAS on Pancreatitis In Vivo

Two experimental models of AP in Wistar rats have been employed to investigate the roles of the RAS in pancreatitis in our laboratory: caerulein-induced AP and obstruction-induced AP. The caerulein-induced AP model, which is well established in our laboratory, garners high reproducibility and specificity. In this model, induction of AP is achieved by administering six successive intraperitoneal injections of caerulein (50 μg/kg/h). Indeed, caerulein-induced severe AP and its associated pulmonary injury have been successfully performed and previously described in our laboratory (Chan & Leung, 2006). The obstructive pancreatitis, or common biliopancreatic duct (CBPD) ligation, model involves opening the abdomen by midline laparotomy and careful externalization of the first loop of the duodenum and a portion of the head of the pancreas. The CBPD is located from the pancreas head extending to the pancreas body, eventually merging with the bile duct. The CBPD is double ligated near the duodenal wall such that the two ligation sites are within 3 cm of each other, thus mimicking a gallstone obstruction, and the abdominal contents are re-internalized (Chan & Leung, 2007a). The animals can be sacrificed at different time points (i.e., 3, 6, 24 h post-ligation) for subsequent experiments. An advantage of the CBPD model over caerulein-induced AP is its high clinical relevance to patients with gallstone obstruction-induced AP (see review by Chan & Leung, 2007b). In this section, all in vivo data are derived from our laboratory (Chan & Leung, 2007a & 2009; Chan, 2008) which will be critically discussed below.

10.2.1 Effects of RAS Blockade in Rat Model of Caerulein-Induced AP

RAS involvement in AP has been examined by experiments testing the effects of AT_1 receptor antagonism on the severity of AP and associated systemic inflammation, particularly in the lungs (Chan & Leung, 2006). Pancreatic oedema, a major AP assessment, is commonly expressed in terms of water content gain (i.e. a ratio of pancreas weight to body weight). Successive injections of caerulein can increase the pancreas to body weight ratio by about threefold. Changes in plasma levels of amylase activity are also indicative of AP diagnosis. Plasma amylase levels are raised sixfold in the caerulein-induced rat model in our laboratory. Our observations of elevated serum amylase and pancreatic to body weight ratio in AP is consistent with previously reported results from other laboratories (Bhatia et al., 2000). Neutrophil sequestration in tissues is a reflection of the degree of AP-induced severity as evidenced by a change in myeloperoxidase (MPO) activity. In this regard, pancreatic MPO activity is elevated 50-fold in our caeruelin-treated animals relative to controls. Interestingly, treatment with the AT_1 receptor antagonist losartan significantly decreases pancreatic MPO activity (fourfold reduction relative to untreated AP group). Histological examination reveals that AP pancreata display severe oedema with large interstitial spaces and a high degree of destruction of the histoarchitecture of the acini. The integrity and architecture of the acini also appear to be improved with losartan treatment (Chan & Leung, 2006). Table 10.1 summarizes the scoring of the histological evaluations of the pancreatic tissue from three independent assessors.

The ability of losartan to affect pathophysiology beyond the pancreas is promising. Lung MPO activity is significantly elevated in caerulein-induced AP animals, increased 5.4-fold relative to controls. We found that losartan significantly ameliorated neutrophil infiltration into the lung tissue, as evidenced by the suppression of pulmonary MPO activity; losartan also attenuated elevated hepatic MPO activity by ~58% (Chan & Leung, 2006). Pulmonary microvascular permeability is an index of macromolecule leakage from the circulation into the alveolar space. After treatment with losartan, pulmonary microvascular permeability in AP (2.6-fold greater than controls) returned to near normal levels (Chan & Leung, 2006). Intact lung tissue morphology is characterized by a discrete alveolar structure, consisting of flattened epithelial cells surrounded by a rich network of pulmonary capillaries and vessels. The alveolar thickening and vasocongestion observed in AP animals was almost normalized in animals injected with losartan (Chan & Leung, 2006). The degree of AP-induced pulmonary injury observed in our study was comparable to that of previous studies (Bhatia et al., 2003; Yamanaka et al., 1997). Table 10.1

Table 10.1 Scoring system for histological evaluation of the pancreatic and pulmonary tissues. All data are obtained from at least three independent experiments and each group of experiment consists of at least 5 individual animals. $^{***}p < 0.001$ vs. AP group; $n = 5$ per group (the data are modified from Chan & Leung, 2006)

Criteria	Control	AP group	AP + losartan group
Pancreatic edema	0±0	1.667±0.126	1.533±0.133
Destruction of pancreatic acinar histoarchitecture	0±0	1.933±0.067	1.200±0.107***
Alevolar thickening	0.733±0.182	1.8±0.145	0.533±0.133
Alevolar vasoconstriction	0.667±0.159	1.467±0.133	0.200±0.106***

summarizes the histological evaluation scores of pulmonary tissue from three independent assessors. These in vivo studies using multiple doses of caerulein-induced AP have demonstrated potential beneficial effects of losartan on AP-induced systemic inflammation. Our experiments involved a more severe AP model than that in previous studies which used a relatively mild model of AP induced by one or two injections of caerulein (Tsang et al., 2003, 2004a, b; Ip et al., 2003b). Taken together, these convergent data provide a clear demonstration that RAS blockade, namely AT_1 receptor antagonism, ameliorates caerulein-induced AP and its associated pulmonary injury. The protective mechanism(s) of AT_1 receptor antagonism in AP-induced inflammation, especially with other clinically relevant animal models, remains to be delineated.

10.2.2 Effects of RAS Blockade in Rat Model of Obstructive AP

The above discussed caerulein-induced AP model has a major disadvantage of weak clinical relevance. The obstruction-induced AP or CBPD ligation model, however, closely mimics the clinical situation of patients with gallstone-obstruction AP (Chan & Leung, 2007a, b). We propose that upregulation of a pancreatic RAS induces locally produced angiotensin II that contributes to the AT_1 receptor-dependent NFκB activation-mediated proinflammatory actions that occur during CBPD ligation-induced AP (Chan & Leung, 2007b). To test this hypothesis, we examined the differential effects of losartan on expression of NFκB and its related proinflammatory genes and proteins in our well-established rat model of obstruction-induced AP.

Real-time polymerase chain reaction (PCR) experiments showed that CBPD ligation time-dependently increased the expression of mRNA of angiotensinogen (the precursor of angiotensin II) at 3, 6, and 24 h post-ligation by approximately 2.1, 2.8, and 4.4-fold, respectively. In corroboration, immunoblot analysis showed 2.7, 3 and 5-fold upregulation of angiotensinogen protein levels at the same time points. AT_1 receptor expression did not differ between AP and control groups at any of the time points studied. However, protein expression of AT_2 receptors was significantly upregulated, by 2.1-fold, at the 6-h time point. We proceeded to test the treatment efficacy of losartan at three dose levels (0.3, 3 and 30 mg/kg) on pancreas to body mass ratio, serum amylase, and pancreatic MPO and IL-6 levels. Ligation of the CBPD significantly elevated pancreas to body mass ratio and serum amylase, 2.4-fold and 4.8-fold of control levels, respectively. The CBPD ligated animals also had a near 190-fold increase in pancreatic MPO activity and a 64-fold elevation in serum IL-6; treatment with losartan (30 mg/kg) significantly reduced these two parameters (by 63 and 60%, respectively) (Chan & Leung, 2007a). Morphological assessment revealed similar effects of the losartan treatment on tissue integrity (Fig. 10.1).

In light of the beneficial effects of losartan on AP-induced inflammation and injury as discussed above, the immediate issue to be addressed is the underlying mechanism(s) of these AT_1 receptor blockade-induced protective effects. To address this issue, the potential regulatory role of the AT_1 receptor in the expression and

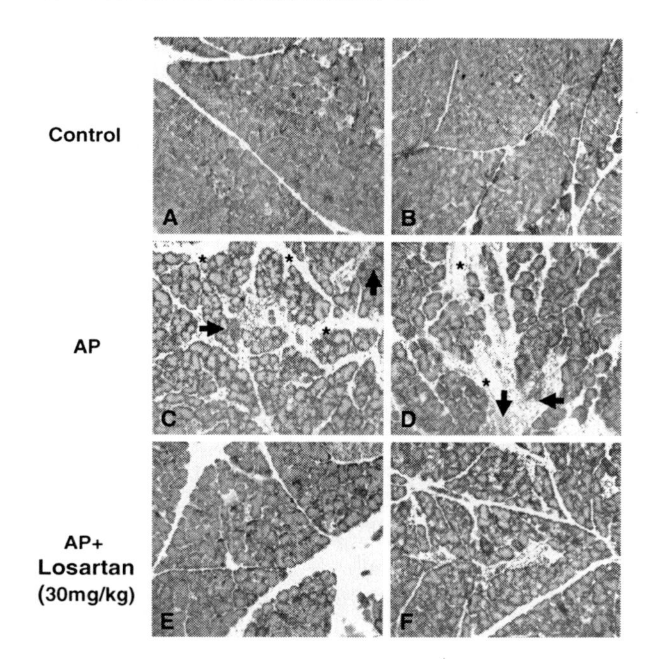

Fig. 10.1 Representative images of pancreatic tissue sections from sham operated controls (**a** and **b**), CBPD-obstruction AP (**c** and **d**) and losartan-treated AP (at 30 mg/kg) (**e** and **f**). Remarkable enlargement in interlobular space (shown by *asterisk*) and inflammatory cell infiltration (shown by *arrow*) was observed in the pancreatic sections in AP group. Results were obtained from at least three separated experiments and each group represents ≤ 5 individual animals. Bar = 80 μm (the data are extracted from Chan & Leung, 2007a)

activity of NADPH oxidase, oxidative stress and NFκB, and κB related proteins that might trigger proinflammatory reactions during AP pathogenesis warrant further investigation. We previously showed that NADPH oxidase activity is augmented in the caerulein-induced AP model (Tsang et al., 2004a). In view of this observation, the expression profile of NADPH oxidase was targeted for further examination in our obstruction-induced AP model. Immunoblot analysis indicated that ligation of CBPD elevated expression of the p67 (a cytoplasmic isoform) and p22 (a membrane-bound isoform) subunits of NADPH oxidase, by 7.2-fold and 3.9-fold, relative to controls levels, respectively. Losartan treatment significantly attenuated the concomitant upregulated expression of the NADPH oxidase subunits according to both immunoblot and immunohistochemistry observations. More importantly, double immunostaining confirmed that the two NADPH oxidase subunits are co-localized with amylase, indicating that the cellular source of NADHP oxidase during AP may originate from pancreatic acinar cells. In addition, glutathione (GSH), the major source of cellular antioxidant in the pancreas, was significantly depleted in the obstructive AP group, while losartan treatment significantly protected against GSH depletion. In keeping with the results of the GSH assay, nitrotyrosine expression (an oxidative stress marker from nitrosylation of protein tyrosine residue) was also significantly elevated in obstructive AP, and losartan treatment significantly attenuated nitrotyrosine expression. Furthermore, double immunohistochemistry indicated that the oxidative stress marker was expressed in acinar cells, suggesting that the oxidative stress molecules arose from pancreatic acinar cells (Chan & Leung, 2007a). All these findings are in close agreement with findings obtained using dihydroethidium (DHE) staining, a fluorescent reactive marker of superoxide free radical. Losartan treatment significantly blunted the formation of fluorescence DHE signal, indicating that ROS generation is likely triggered by AT_1 receptor activation (Chan & Leung, 2009).

It remains to be resolved whether ROS generation is linked to NFκB activation so as to exert the pro-inflammatory actions. In order to study the participation of NFκB-induced pancreatic inflammation in AP, the expression profile of NFκB, in terms of its inhibitory subunits IκB (IκBα an IκBβ), activated form of NFκB (phospho-NFκB p65), and nuclear binding activity needs to be delineated. Our data show that CBPD ligation results in degradation of IκBβ, an effect which is largely reversed by losartan treatment without affecting IκBα levels (Chan & Leung, 2007a). Concomitantly, phospho-NFκB p65 protein levels are upregulated 16-fold during AP, but are significantly suppressed after losartan treatment. In addition, electrophoretic mobility shift assay (EMSA) further showed that nuclear extracts from the pancreata of obstructive AP animals exhibit a diminished electrophoretic mobility upon electrophoresis compared with control group extracts, indicating that there is enhanced nuclear binding activity of the κB sequence in the AP condition. Meanwhile, losartan treatment also significantly reversed this nuclear κB binding activity (Chan & Leung, 2007a).

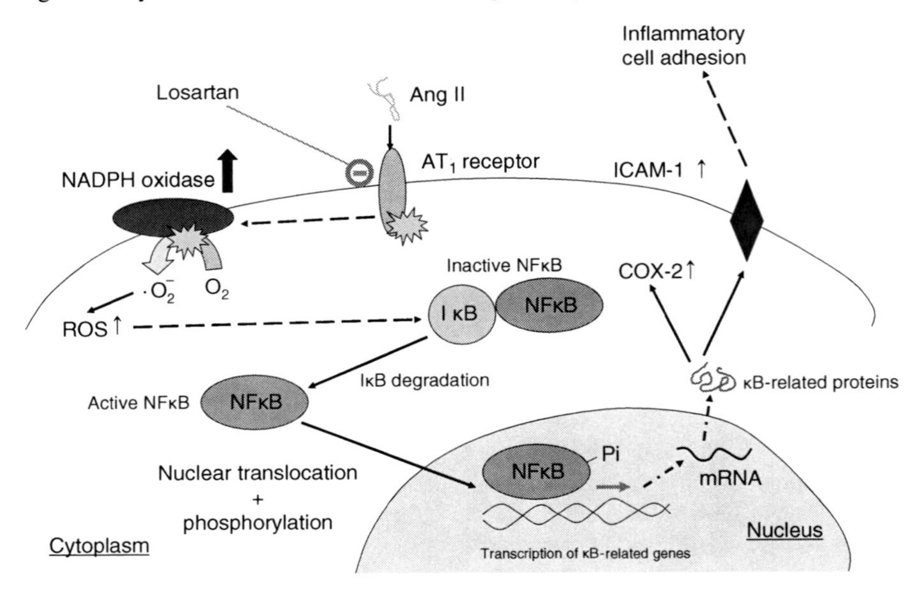

Fig. 10.2 A schematic diagram illustrating the putative mechanisms of the AT_1 receptor-ROS-NFκB axis involved in the regulation of proinflammatory signaling cascade during the pathogenesis of pancreatitis (modified from Chan, 2008)

NFκB elicits its inflammatory actions by means of transcription of an array of downstream mediators, including intracellular adhesion molecule (ICAM)-1, the prostaglandin-producing enzyme cyclooxygenase (COX)-2, and the pro-inflammatory cytokine interleukin (IL)-1α. In this regard, immunoblot results revealed that CBPD obstruction enhances ICAM-1 and COX-2 protein expression by 7 and 22-fold, respectively; this upregulation of κB-related proteins was abolished by losartan treatment (Chan & Leung, 2007a). Likewise, serum IL-1 was augmented in the obstructive AP group and this effect was significantly attenuated by losartan treatment (Chan & Leung, 2007a). Taken together, the above reviewed data provide substantial evidence that pancreatic AT_1 receptor is involved in NADPH oxidase-dependent NFκB activation-induced proinflammatory actions in pancreatic inflammation. Figure 10.2 summarizes the pathways proposed to be involved in this AT_1 receptor-ROS-NFκB axis in the regulation of the proinflammatory cascade during pancreatic inflammation.

10.3 Current Research of the RAS on Pancreatitis In Vitro

In addition to switching on the NFκB transduction pathway, ROS also activate the redox-sensitive mitogen-activated protein kinase (MAPK) in multiple cell types. The MAPKs, in particular, extracellular-regulated kinase (ERK)1/2, are responsive to a variety of stimuli including cytokines (notably interleukin-6, IL-6), growth factors, ROS, and cellular stress, and thus ultimately govern a variety of cellular processes including cytoskeleton arrangement, transcription factor activation, apoptosis, proliferation and differentiation (Leung & Chan, 2009). Interestingly, angiotensin II-mediated ROS production induces activation of ERK1/2 in several cell types such as renal proximal tubule cells (Tanifuji et al., 2005), vascular smooth muscle cells (Viedt et al., 2000), and neutrophils (El Bekay et al., 2003). Intriguingly, ERK1/2-mediated IL-6 expression is also dependent on the participation of other nuclear proteins as ERK1/2 is not a direct transcription factor. ERK1/2 has been reported to control many downstream mediators to ultimately influence the transcription of various genes. Among them, cAMP-responsive element binding protein (CREB) is one of candidates proposed to be closely associated with inflammatory response. It has been shown that the promoter of a number of pro-inflammatory genes, including COX-2 and IL-6, contains the cAMP-responsive element (CRE) (Ichiki, 2006). Targeted disruption of the CRE site of the COX-2/IL-6 promoter impairs stimulus-induced pro-inflammatory gene expression, suggesting that CREB may be able to regulate inflammatory responses (Ichiki, 2006; Sano et al., 2001). These convergent findings, notwithstanding the specific cell type within the pancreas involved in angiotensin II-induced redox-regulated pathways during pancreatic inflammation, have yet to be determined. Addressing this issue will require employing an in vitro system of a pancreatic acinar cell line, namely AR42J, to elaborate the potential participation of redox-sensitive ERK1/2, with particular focus on angiotensin II-induced IL-6 expression in the regulation of pancreatic acinar cells during pancreatic inflammation. Ongoing investigations employing the current standard approaches will provide substantial information about the intricate interactions between the RAS and ROS, and the RAS' concomitant signalling pathways in the genesis of pancreatic inflammation. In this section, all in vitro data are derived from our laboratory (Chan & Leung, 2007a & 2009; Chan, 2008) which will be critically discussed below.

10.3.1 Effects of Angiotensin II on ERK and CREB-Mediated IL6 Expression

As a first step towards achieving the above objectives, angiotensin II-induced ERK and CREB activation of IL-6 are being examined. Results thus far have shown that exogenous addition of angiotensin II augments ERK1/2 activation in a dose- and time-dependent manner, such that peak ERK1/2 activation is seen with 1 μM dose of angiotensin II at the 5-min time point. Similarly, application of angiotensin II also dose and time dependently enhances IL-6 protein expression, with the peak effect observable with a 1 μM, at 6 h post-application. Exogenous administration of 1 μM angiotensin II to AR42J cells also leads to activation of CREB expression 1 h post-treatment (Chan & Leung, 2009). These available data have prompted us to propose that angiotensin II may play an active role in ROS-mediated ERK-induced CREB activation and subsequent IL-6 expression, thus triggering inflammation in pancreatic acinar cells (see review by Leung & Chan, 2009).

Most, if not all, of the effects of angiotensin II on pro-inflammatory pathways act through AT1R in cells and tissues. Losartan treatment of acinar cells at 50 μM, but not in the range of 0.5–5 μM, effects ERK1/2 activation within 5 min. This AT1R-mediated ERK activation was confirmed by negative control pre-treatment of the cells with PD98059, an ERK inhibitor (5 μM). The effects of different blockers on CREB activation have also been examined. Pre-treatment of AR42J cells with losartan (50 μM) and PD98059 (5 μM) can significantly abolish the action of angiotensin II-induced CREB activation, suggesting that ERK may be an upstream mediator of CREB activation. Likewise, pre-treatment of losartan (50 μM) significantly abolishes the angiotensin II-induced IL-6 protein expression. More interestingly, angiotensin II-induced IL-6 expression can be blocked by the specific ERK inhibitor PD98059 (5 μM), thus implying that angiotensin II-induced IL-6 expression depends, at least partly, on the ERK activation (Chan & Leung, 2009).

10.3.2 Potential Crosstalk Between Angiotensin II-Induced ROS-Mediated ERK and NFκB

To further explore whether there is an interaction between angiotensin II-dependent ROS-induced ERK and NFκB activation in the regulation of proinflammatory pathways during pancreatitis, the time-dependent effects of exogenous angiotensin II

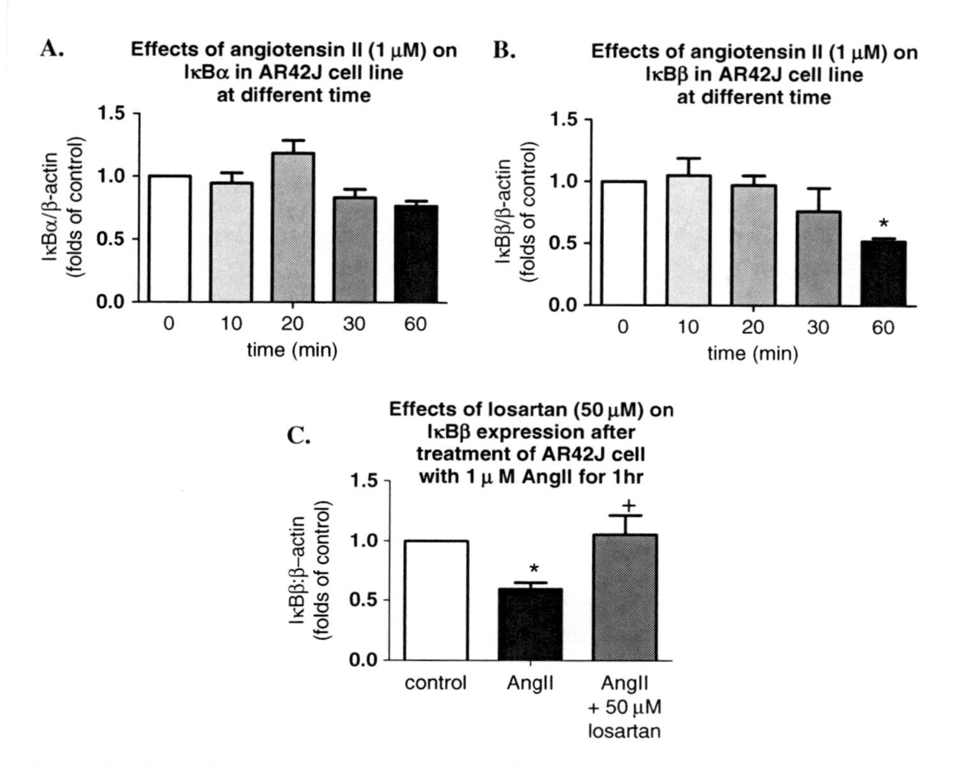

Fig. 10.3 Effects of angiotensin II on protein expression of IκBα (**a**) and IκBβ (**b**), and losartan on protein expression of IκBβ (**c**) in AR42J cell line. * indicates $p < 0.05$ vs. control while $^{+}$ represent $p < 0.05$ vs. angiotensin II alone. Results were obtained from at least three separated experiments with different passage number (Chan & Leung, published data)

on the expression of NFκB inhibitory subunits were investigated in the AR42J cell line (Chan & Leung, unpublished data). Immunoblot analysis revealed that IκBα expression was not altered by stimulation with angiotensin II (1 μM) at any of the time points studied (i.e. from 10 to 60 min). Instead, treatment of the acinar cell line with angiotensin II (1 μM) triggered degradation of IκBβ which reach about 50% by 60 min post-treatment; pre-treatment with losartan prevented this degradation (Fig. 10.3). These findings are consistent with our in vivo results (see Section 10.2) in suggesting that acinar cells are responsive to angiotensin II-induced activation of NFκB.

On the other hand, a previous study demonstrated that ERK1/2 can phosphorylate the IκB kinase (IKK) complex, resulting in NFκB activation (Liu & Wong, 2005). It has also been demonstrated that ERK is required for activation of NFκB in cultured rat vascular smooth muscle cells (VSMCs) (Jiang et al., 2004). Inhibition of ERK can lead to resistance of IL-1β-induced IκBβ degradation, indicative of the ERK involvement in temporal regulation of NFκB activation (Jiang et al., 2004).

These findings suggest that ERK might serve as an upstream mediator of NFκB via activation of the IKK complex in pancreatic acinar cells. It is thus plausible to speculate that ERK may be responsible for angiotensin II-induced NFκB activation. In order to address this issue, we tried to pre-treat pancreatic acinar cells with an ERK inhibitor (PD98059); as such, it did not exert any effects on angiotensin II-induced IκBβ degradation. On contrary, pre-treatment of the AR42J cells with DPI (10 μM), an NADPH oxidase inhibitor can markedly decrease angiotensin II-mediated IκBβ degradation (about 90% of the control). These data suggest that angiotensin II-induced NFκB activation in pancreatic acinar cells involves NADPH oxidase, but not ERK1/2 (Fig. 10.4a). In order to study the potential role of ROS in ERK1/2 activation, angiotensin II-stimulated AR42J cells were subjected to DHE staining and examined. Pre-treatment of the pancreatic acinar cells with losartan, but not PD123319, markedly abolished ROS generation induced by exogenous application of angiotensin II (Chan & Leung, 2009). These observations indicate that angiotensin II may stimulate ROS production in an AT_1 receptor-mediated manner. More importantly, activation of ERK1/2 was significantly reduced by pretreatment with the antioxidant NAC, indicating that ERK1/2 is subject to redox-regulation in pancreatic acinar cells (Fig. 10.4b).

Finally, we examined the potential role of NFκB in IL-6 expression. In order to achieve this, we did pre-treat AR42J cells with an NADPH oxidase inhibitor (DPI) or an NFκB inhibitor (PDTC); by doing so, it prevented angiotensin II-mediated induction of IL-6 expression (Fig. 10.5). These findings indicate that angiotensin II robustly affects IL-6 transcription through induction of NADPH oxidase and subsequent activation of NFκB, which are key mediators in eliciting IL-6 transcription. Taken all the data together, Fig. 10.6 provides a schematic overview of the potential crosstalk of the AT1R-ROS-ERK1/2-NFκB axis in the regulation of proinflammatory signalling during pancreatic inflammation.

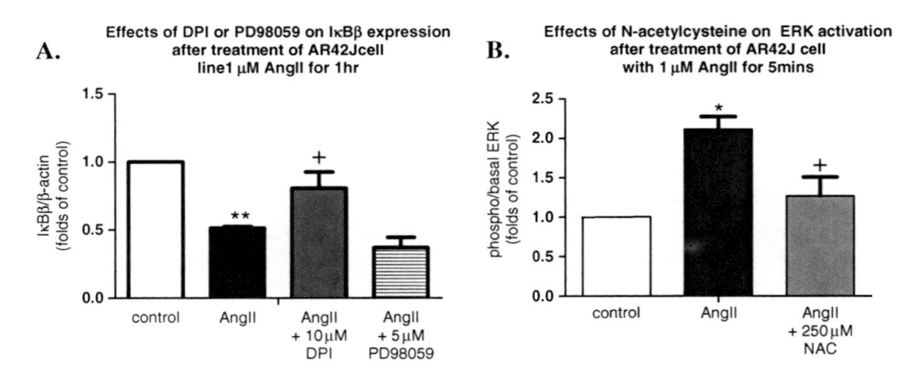

Fig. 10.4 **a** Effects of NADPH oxidase and ERK inhibitor on IκBβ protein expression in AR42J cell line after treatment of angiotensin II for 1 h. **b** Effects of antioxidant on on protein expression of phospho ERK in AR42J cell line after treatment of angiotensin for 5 min. * indicates $p< 0.05$ and ** $p<0.01$ vs. control while $^+$ indicates $p< 0.05$ vs. angiotensin II only (Chan & Leung, 2009)

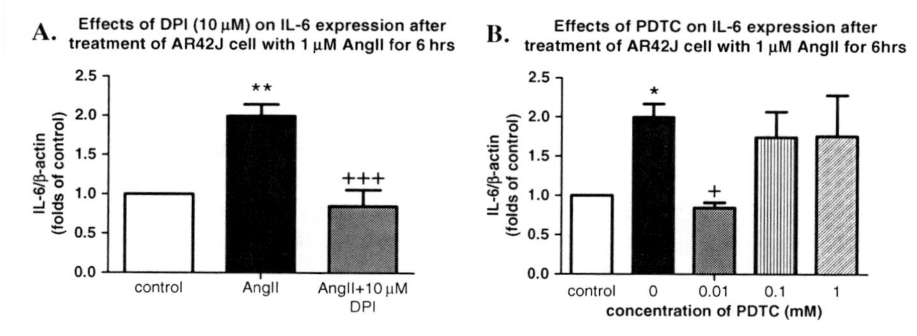

Fig. 10.5 Effects of NADPH oxidase (**a**) and NFκB inhibitor (**b**) on protein expression of IL-6 in AR42J cell line after treatment of angiotensin II for 6 h. ** indicates $p < 0.01$, $^{*}p<0.05$ vs. control while $^{+++}$ indicates $p < 0.001$, $^{+}p<0.05$ vs. angiotensin II only (Chan & Leung, unpublished data)

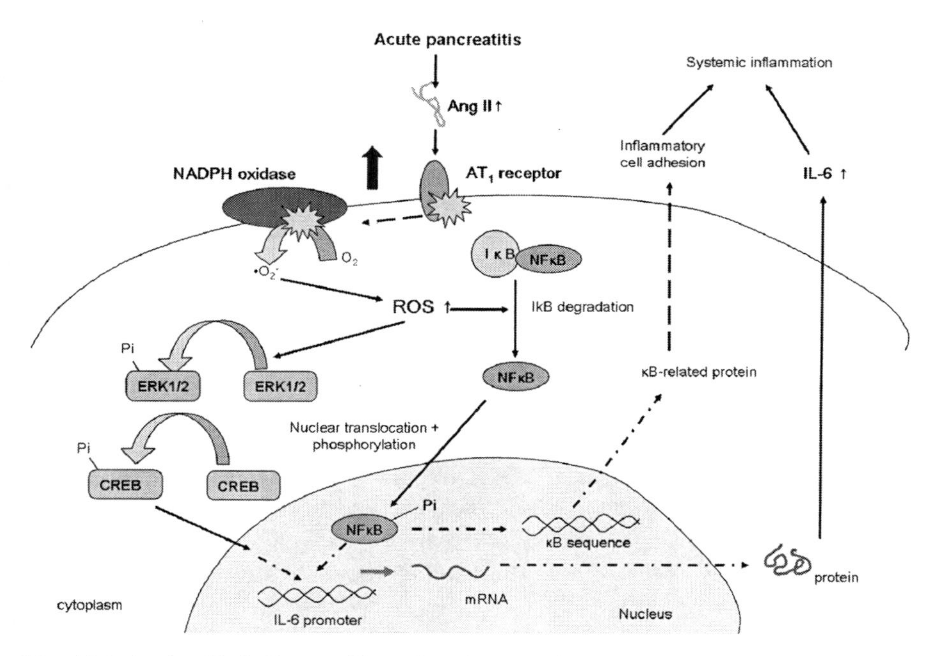

Fig. 10.6 A simplified diagram illustrating the potential crosstalk between AT_1 receptor-dependent ROS-mediated ERK1/2 and NFκB activation thus IL-6 induced proinflammatory signaling cascade during the pathogenesis of pancreatitis (modified from Chan, 2008)

10.4 Current Research of the RAS in Pancreatic Cancer

The second part of this Chapter provides a critical review of current progress in research concerning the RAS and pancreatic cancer. Pancreatic cancer is generally divided into two types: pancreatic exocrine tumour and pancreatic endocrine tumour

(PET). About 95% of exocrine pancreatic cancers are of pancreatic ductal adenocarcinomas (PDACs), which are far more common than PETs, the latter constituting only 1% of the total cases of pancreatic cancer (See Table. 3.1, Chapter 3). As discussed earlier in this book (Chapters 4 and 5), the RAS plays endocrine, paracrine, autocrine and intracrine roles in the regulation of various disease-associated cell functions. In this regard, the RAS acts as a potent growth factor not only for VSMCs, but also for certain cancer cells (Fujimoto et al., 2001; Muscella et al., 2002). Angiotensin II stimulates angiogenesis by upregulating vascular endothelial growth factor (VEGF) (Tamarat et al., 2002). Interestingly, the RAS is subject to activation by neoplastic conditions (Takeda & Kondo, 2001; Juillerat-Jeanneret et al., 2004).

10.4.1 Expression and Regulation of the RAS Components in Tumours

In tumours, angiotensin II is abundantly produced by overactive ACE activity while AT_1 receptor expression is significantly upregulated. Previous studies in experimental animal models have shown that the RAS is intimately involved in tumour growth, metastasis and angiogenesis, suggesting that pharmacological RAS blockade may have therapeutic potential for some forms of cancer (Rivera et al., 2001; Uemura et al., 2003; Arrieta et al., 2005). Several RAS components have been observed in various cancers including, to name but a few, brain, lung, breast, prostate, skin and pancreatic cancer (Deshayes & Namias, 2005).

Angiotensinogen and AT_2 receptor are both upregulated in rat PET cells and human PET tissues (Lam, 2001; Lam & Leung, 2002). High concentrations of angiotensin II and upregulated expression of AT_1 receptor have been described in human PDAC cell lines and tissues (Fujimoto et al., 2001; Arafat et al., 2007; Anandanadesan et al., 2008). Furthermore, ACE and AT_1 receptor are overexpressed in the majority of cases of PDAC in human patients, and ACE plays a crucial role in the occurrence of PDAC through the formation of angiotensin II (Arafat et al., 2007). Inhibition of ACE modulates mitosis in pancreatic cancer cells (Arrieta et al., 2005; Egami et al., 2003). Indeed, ACE and angiotensin II levels are higher in PDAC tissues than in adjacent tumour-free tissues. In contrast, ACE2 protein expression is lower in PDAC tissues than in tumour-free tissues. ACE2, a homologue of ACE, converts angiotensin II into angiotensin (1–7), another bioactive peptide of the RAS. Angiotensin (1–7) acts as a vasodilator and plays a role in apoptosis and cell growth, thus opposing the actions of angiotensin II (see Chapter 4). ACE2, which is highly expressed in the endothelial cells of the heart, kidney and testis, appears to counterbalance the angiotensin II-promoting effect of ACE and, presumably, plays a protective role in cardiometabolic diseases (Raizada & Ferreira, 2007; Mizuiri et al., 2008; Wang et al., 2008). Of particular interest in this context is the decreased expression of ACE2 protein in the presence of angiotensin II in pancreatic cancer cells (Zhou et al., 2009). All of these convergent data thus implicate that the RAS may be involved in the regulation of tumor cell proliferation, angiogenesis and migration in patients with PDAC.

It is now widely accepted that the growth of solid tumours depends on angiogenesis. Cancer researchers worldwide are working to overcome tumour angiogenesis (Hanahan & Weinberg, 2000). Nevertheless, angiography in the clinic reveals that most pancreatic cancers are hypovascular or avascular, though human pancreatic cancer cells have been reported to overexpress pro-angiogenic factors such as VEGF (Kuwahara et al., 2003). In animal models, anti-angiogenic compounds have been shown to inhibit pancreatic tumour development (Laquente et al., 2008; Zhang et al., 2005). On the other hand, PDAC exhibits foci of microangiogenesis and overexpresses several pro-angiogenic factors. VEGF represents a crucial factor in the pro-angiogenic switch that occurs in PDAC (Korc, 2001). In the normal pancreas, VEGF expression is limited to the islets, co-localizing with insulin, and, interestingly, is inducible by hypoglycemia (Kuroda et al., 1995). In PDAC, high VEGF and VEGF receptor levels are associated with enhanced lymph node metastasis, poor prognosis, and postoperative recurrence (Büchler et al., 2000). VEGF itself has also been implicated as a survival factor for certain tumour cells, rendering them radioresistant (Gupta et al., 2002).

10.4.2 RAS Blockade and Pancreatic Cancers

Angiotensin II was recently shown to be involved in the regulation of cell proliferation, migration, inflammation and tissue remodelling as well as angiogenesis via AT_1 receptor. Angiotensin II stimulates in vitro cell proliferation, invasion, and VEGF secretion via AT_1 receptor in PDAC. In this context, it has been reported that human pancreatic cancer cells inappropriately express AT_1 receptor and angiotensin II, enhancing tumour cell invasion and VEGF expression while AT_1 receptor antagonism suppresses angiogenesis in pancreatic cancer (Amaya et al., 2004; Anandanadesan et al., 2008). In fact, other studies have shown that ACE and AT_1 receptor blockers could help reduce angiogenesis in PDAC owing to their strong anti-angiogenic activity, thus inhibiting tumour growth and lowering the incidence of cancer development (Yoshiji et al., 2002). It has been reported that a selective AT_1 receptor antagonist suppresses growth of human pancreatic cancer cells in vitro (Fujimoto et al., 2001). The ability of ACE and AT_1 receptor blockers to ameliorate angiotensin II actions and production suggest that they may be useful in the treatment and management of pancreatic cancer. Indeed, RAS antagonism inhibits the growth of pancreatic cancer cells in vitro (Fujimoto et al., 2001). Recent studies have further demonstrated that expression of ACE2 is enhanced with inhibition of ACE and AT_1 receptor. It is also noteworthy that there is an imbalance of the ACE/ACE2 ratio in PDAC, suggesting a possible role of ACE2 in PDAC (Ocaranza et al., 2006; Igase et al., 2008).

ACE and AT_1 receptor protein expression was found to be elevated in 75% of examined pancreatic cancer specimens, and VEGF expression was also higher in cases of elevated ACE and AT_1 receptor levels. Both ACE and AT_1 receptor are functionally expressed in PDAC and inhibition of the RAS enhances the VEGF-mediated angiogenic process in PDAC. Furthermore, angiotensin II

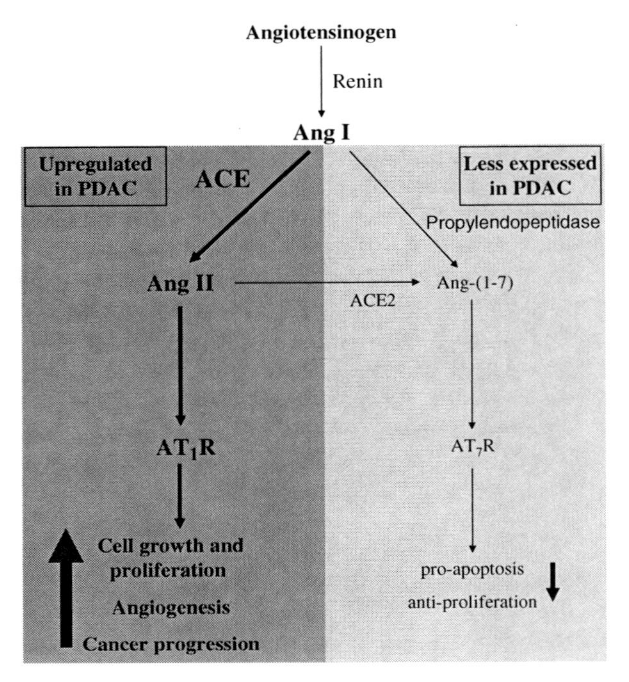

Fig. 10.7 A schematic representation of proposed roles of the RAS, with particular emphasis on AT1R (AT_1 receptor) and AT7R (AT_7 receptor or Mas), in the regulation of tumour growth, angiogenesis and metastasis observed in patients with PDAC

enhances VEGF production in AT_1 receptor-positive cells in vitro; and treatments with an ACE inhibitor (captopril) and AT_1 receptor antagonist (losartan) are able to block these effects. Indeed, oral administration of captopril may protect against cancer (Lever et al., 1998) and its use in combination with other anti-angiogenic agents has been tested in a phase II study (Deplanque & Harris, 2000). Considering these findings, we posit that angiotensin II may act as an angiogenic and tumour-progressive factor for pancreatic cancer. We further suggest that the expression profile of AT_1 receptor make it a molecular target or prognostic indicator for pancreatic cancer treatment. Thus, pharmacological intervention of the RAS may offer an alternative to current courses of pancreatic cancer treatment and management (see recent review by Lau & Leung, 2010). Figure 10.7 provides a summary of the putative effects of angiotensin II and Ang (1–7) in the regulation of cell growth and proliferation of human PDAC.

References

Alvarez A and Sanz MJ. Reactive oxygen species mediate angiotensin II-induced leukocyte-endothelial cell interactions in vivo. *J Leukoc Biol* **70**:199–206, 2001.

Amaya K, Ohta T, Kitagawa H, Kayahara M, Takamura H, Fujimura T, Nishimura G, Shimizu K and Miwa K. Angiotensin II activates MAP kinase and NF-kappaB through angiotensin II type I receptor in human pancreatic cancer cells. *Int J Oncol* **25**:849–856, 2004.

Anandanadesan R, Gong Q, Chipitsyna G, Witkiewicz A, Yeo CJ and Arafat HA. Angiotensin II induces vascular endothelial growth factor in pancreatic cancer cells through an angiotensin II type 1 receptor and ERK1/2 signaling. *J Gastrointest Surg* **12**:57–66, 2008.

Arafat HA, Gong Q, Chipitsyna G, Rizvi A, Saa CT and Yeo CJ. Antihypertensives as novel antineoplastics: angiotensin-I-converting enzyme inhibitors and angiotensin II type 1 receptor blockers in pancreatic ductal adenocarcinoma. *J Am Coll Surg* **204**:996–1005, 2007.

Arrieta O, Guevara P, Escobar E, García-Navarrete R, Pineda B and Sotelo J. Blockage of angiotensin II type I receptor decreases the synthesis of growth factors and induces apoptosis in C6 cultured cells and C6 rat glioma. *Br J Cancer* **92**:1247–1252, 2005.

Babior BM. NADPH oxidase: an update. *Blood* **93**:1464–1476, 1999.

Beckman JS and Koppenol WH. Nitric oxide, superoxide, and peroxynitrite: the good, the bad, and ugly. *Am J Physiol* **271**:C1424–C1437, 1996.

Bendall JK, Cave AC, Heymes C, Gall N and Shah AM. Pivotal role of a gp91(phox)-containing NADPH oxidase in angiotensin II-induced cardiac hypertrophy in mice. *Circulation* **105**: 293–296, 2002.

Bhatia M, Slavin J, Cao Y, Basbaum AI and Neoptolemos JP. Preprotachykinin-A gene deletion protects mice against acute pancreatitis and associated lung injury. *Am J Physiol* **284**: G830–G836, 2003.

Bhatia M, Brady M, Zagorski J, Christmas SE, Campbell F, Neoptolemos JP and Slavin J. Treatment with neutralising antibody against cytokine induced neutrophil chemoattractant (CINC) protects rats against acute pancreatitis associated lung injury. *Gut* **47**:838–844, 2000.

Büchler P, Reber HA, Büchler MW, Friess H and Hines OJ. VEGF-RII influences the prognosis of pancreatic cancer. *Ann Surg* **236**:738–749, 2002.

Chan YC and Leung PS. Involvement of redox-sensitive extracellular-regulated kinases in angiotensin II-induced interleukin-6 expression in pancreatic acinar cells. *J Pharmacol Exp Ther* **329**:450–458, 2009.

Chan YC. The novel role of angiotensin II in acute pancreatitis. PhD thesis, The Chinese University of Hong Kong, 2008.

Chan YC and Leung PS. Angiotensin II type 1 receptor-dependent nuclear factor-κB activation-mediated proinflammatory actions in a rat model of obstructive acute pancreatitis. *J Pharmacol Exp Ther* **323**:10–18, 2007a.

Chan YC and Leung PS. Acute pancreatitis: animal models and recent advances in basic research. *Pancreas* **34**:1–14, 2007b.

Chan YC and Leung PS. AT_1 receptor antagonism ameliorates acute pancreatitis-associated pulmonary injury. *Regul Pept* **134**:46–53, 2006.

Chan WP, Fung ML, Nobiling R and Leung PS. Activation of local renin-angiotensin system by chronic hypoxia in rat pancreas. *Mol Cell Endocrinol* **160**:107–114, 2000.

Chappell MC, Millsted A, Diz DI, Brosnihan Kb and Ferrario CM. Evidence for an intrinsic RAS in the canine pancreas. *J Hypertens* **9**:751–759, 1991.

Chappell MC, Diz DI and Jacobsen DW. Pharmacological characterization of Ang II binding sites in pancreas. *Peptides* **14**:311–318, 1992.

Czakó L, Takács T, Varga IS, Tiszlavicz L, Hai DQ, Hegyi P, Matkovics B and Lonovics J. Oxidative stress in distant organs and the effects of allopurinol during experimental acute pancreatitis. *Int J Pancreatol* **27**:209–216, 2000.

Dang PM, Morel F, Gougerot-Pocidalo MA and El Benna J. Phosphorylation of the NADPH oxidase component p67(PHOX) by ERK2 and P38MAPK: selectivity of phosphorylated sites and existence of an intramolecular regulatory domain in the tetratricopeptide-rich region. *Biochemistry* **42**:4520–4526, 2003.

De Gasparo M and Siragy HM. The AT_2 receptor: fact, fancy and fantasy. *Regul Pept* **81**:11–24, 1999.

De Gasparo M, Catt KJ, Inagami T, Wright JW and Unger T. The angiotensin II receptors. *Pharmacol Rev* **52**:415–472, 2000.

De Gasparo M. Angiotensin II and nitric oxide interaction. *Heart Fail Rev* **7**:347–358, 2002.

Deplanque G and Harris AL. Anti-angiogenic agents: clinical trial design and therapies in development. *Eur J Cancer* **36**:1713–1724, 2000.

Deshayes F and Nahmias C. Angiotensin receptors: a new role in cancer? *Trends Endocrinol Metab* **16**:293–299, 2005.

Dijkhorst-Oei LT, Stroes ES, Koomans HA and Rabelink TJ. Acute simultaneous stimulation of nitric oxide and oxygen radicals by angiotensin II in humans in vivo. *J Cardiovasc Pharmacol* **33**:420–424, 1999.

Egami K, Murohara T, Shimada T, Sasaki K, Shintani S, Sugaya T, Ishii M, Akagi T, Ikeda H, Matsuishi T and Imaizumi T. Role of host angiotensin II type 1 receptor in tumor angiogenesis and growth. *J Clin Invest* **112**:67–75, 2003.

El Bekay R, Alvarez M, Monteseirín J, Alba G, Chacón P, Vega A, Martin-Nieto J, Jiménez J, Pintado E, Bedoya FJ and Sobrino F. Oxidative stress is a critical mediator of the angiotensin II signal in human neutrophils: involvement of mitogen-activated protein kinase, calcineurin, and the transcription factor NF-kappaB. *Blood* **102**:662–671, 2003.

Ferreri NR, Escalante BA, Zhao Y, An SJ and McGiff JC. Angiotensin II induces TNF production by the thick ascending limb: functional implications. *Am J Physiol* **274**:F148–F155, 1998.

Folch-Puy E, García-Movtero A, Iovanna JL, Dagorn JC, Prats N, Vaccaro MI and Closa D. The pancreatitis-associated protein induces lung inflammation in the rat through activation of TNFalpha expression in hepatocytes. *J Pathol* **199**:398–408, 2003.

Fujimoto Y, Sasaki T, Tsuchida A and Chayama K. Angiotensin II type 1 receptor expression in human pancreatic cancer and growth inhibition by angiotensin II type 1 receptor antagonist. *FEBS Lett* **495**:197–200, 2001.

Ghiani BU and Masini MA. Angiotensin II binding sites in the rat pancreas and their modulation after sodium loading and depletion. *Comp Biochem Physiol* **111A**:439–444, 1995.

Granell S, Gironella M, Bulbena O, Panés J, Mauri M, Sabater L, Aparisi L, Gelpí E and Closa D. Heparin mobilizes xanthine oxidase and induces lung inflammation in acute pancreatitis. *Crit Care Med* **31**:525–530, 2003.

Grewal HP, Mohey el Din A, Gaber L, Kotb M and Gaber AO. Amelioration of the physiologic and biochemical changes of acute pancreatitis using anti-TNP-alpha polyclonal antibody. *Am J Surg* **167**:214–219, 1994.

Griendling KK, Sorescu D and Ushio-Fukai M. NADPH oxidase: role in cardiovascular biology and disease. *Circ Res* **86**:494–501, 2000.

Gupta VK, Jaskowiak NT, Beckett MA, Mauceri HJ, Grunstein J, Johnson RS, Calvin DA, Nodzenski E, Pejovic M, Kufe DW, Posner MC and Weichselbaum RR. Vascular endothelial growth factor enhances endothelial cell survival and tumor radioresistance. *Cancer J* **8**:47–54, 2002.

Hanahan D and Weinberg RA. The hallmarks of cancer. *Cell* **100**:57–70, 2000.

Heath DI, Cruickshank A, Gudgeon M, Jehanli A, Shenkin A and Imrie CW. Role of interleukin-6 in mediating the acute phase protein response and potential as an early means of severity assessment in acute pancreatitis. *Gut* **34**:41–45, 1993.

Ichiki T. Role of cAMP response element binding protein in cardiovascular remodeling: good, bad, or both? *Arterioscler Thromb Vasc Biol* **26**:449–455, 2006.

Igase M, Kohara K, Nagai T, Miki T and Ferrario CM. Increased expression of angiotensin converting enzyme 2 in conjunction with reduction of neointima by angiotensin II type 1 receptor blockade. *Hypertens Res* **31**:553–559, 2008.

Ip SP, Chan YW and Leung PS. Effects of chronic hypoxia on the circulating and pancreatic RAS. *Pancreas* **25**:296–300, 2002.

Ip SP, Kwan PC, Williams CH, Pang S, Hooper NM and Leung PS. Changes of angiotensin-converting enzyme activity in the pancreas of chronic hypoxia and acute pancreatitis. *Int J Biochem Cell Biol* **35**:944–954, 2003a.

Ip SP, Tsang SW, Wong TP, Che CT and Leung PS. Saralasin, a non-specific angiotensin II receptor antagonist, attenuates oxidative stress and tissue injury in cerulein-induced acute pancreatitis. *Pancreas* **26**:224–229, 2003b.

Jaimes EA, Galceran JM and Raij L. Ang II induces superoxide anion production by mesangial cells. *Kidney Int* **54**:775–784, 1998.

Jiang B, Xu S, Hou X, Pimentel DR, Brecher P and Cohen RA. Temporal control of NF-kappaB activation by ERK differentially regulates interleukin-1beta-induced gene expression. *J Biol Chem* **279**:1323–1329, 2004.

Juillerat-Jeanneret L Celerier J, Chapuis Bernasconi C, Nguyen G, Wostl W, Maerki HP Janzer RC, Corvol P and Gasc JM. Renin and angiotensinogen expression and functions in growth and apoptosis of human glioblastoma. *Br J Cancer* **90**:1059–1068, 2004.

Kalra D, Sivasubramanian N and Mann DL. Angiotensin II induces tumor necrosis factor biosynthesis in the adult mammalian heart through a protein kinase C-dependent pathway. *Circulation* **105**:2198–2205, 2002.

Kelm M, Dahmann R, Wink D and Feelisch M. The nitric oxide/superoxide assay. *J Biol Chem* **272**:9922–9932, 1997.

Kintscher U, Wakino S, Kim S, Fleck E, Hsueh WA and Law RE. Angiotensin II induces migration and Pyk2/paxillin phosphorylation of human monocytes. *Hypertension* **37**:587–593, 2001.

Knoefel WT, Kollias N, Warshaw AL, Waldner H, Nishioka NS and Rattner DW. Pancreatic microcirculatory changes in experimental pancreatitis of graded severity in rat. *Surgery* **116**:904–913, 1994.

Korc M. Pathways for aberrant angiogenesis in pancreatic cancer. *Mol Cancer* **7**:2–8, 2001.

Kuno A, Yamada T, Masuda K, Ogawa K, Sogawa M, Nakamura S, Nakazawa T, Ohara H, Nomura T, Joh T, Shirai T and Itoh M. Angiotensin-converting enzyme inhibitor attenuates pancreatic inflammation and fibrosis in male Wistar Bonn/Kobori rats. *Gastroenterology* **124**:1010–1019, 2003.

Kuroda M, Oka T, Oka Y, Yamochi T, Ohtsubo K, Mori S, Watanabe T, Machinami R and Ohnishi S. Colocalization of vascular endothelial growth factor (vascular permeability factor) and insulin in pancreatic islet cells. *J Clin Endocrinol Metab* **80**:3196–3200, 1995.

Kuwahara K, Sasaki T, Kuwada Y, Murakami M, Yamasaki S and Chayama K. Expressions of angiogenic factors in pancreatic ductal carcinoma: a correlative study with clinicopathologic parameters and patient survival. *Pancreas* **26**:344–349, 2003.

Lam KY. The pancreatic renin-angiotensin system: does it play a role in endocrine oncology. *J Pancreas* **2**:40–42, 2001.

Lam KY and Leung PS. Regulation and expression of a renin-angiotensin system in human pancreas and pancreatic endocrine tumours. *Eur J Endocrinol* **146**:567–572, 2002.

Laquente B, Lacasa C, Ginesta MM, Casanovas O, Figueras A, Galan M, Ribas IG, Germa JR, Capella G and Vinals F. Antiangiogenic effect of gemcitabine following metronomic administration in a pancreas cancer model. *Mol Cancer Ther* **7**:638–647, 2008.

Lau ST and Leung PS. Role of the RAS in pancreatic cancer. *Curr Cancer Drug Targets* (in press), 2010.

Leung PS, Chan HC, Fu LX and Wong PY. Localization of AT_1 and AT_2 receptors in the pancreas of rodents. *J Endocrinol* **153**:269–274, 1997.

Leung PS, Chan HC and Wong PY. Immunohistochemical localization of Ang II in the mouse pancreas. *Histochem J* **30**:21–25, 1998.

Leung PS, Chan WP, Wong TP and Sernia C. Expression and localization of the RAS in the rat pancreas. *J Endocrinol* **160**:13–19, 1999.

Leung PS, Chan WP and Nobiling R. Regulated expression of pancreatic RAS in experimental pancreatitis. *Mol Cell Endocrinol* **166**:121–128, 2000.

Leung PS and Carlsson PO. Tissue renin-angiotensin system: its expression, localization, regulation and potential role in the pancreas. *J Mol Endocrinol* **26**:155–163, 2001.

Leung PS. Pancreatic RAS: a novel target for potential treatment of pancreatic diseases. *J Pancreas* **4**:89–91, 2003.

Leung PS and Chappell MC. A local pancreatic renin-angiotensin system: endocrine and exocrine roles. *Int J Biochem Cell Biol* **35**:834–846, 2003.

Leung PS. Roles of the renin-angiotensin system and its blockade in pancreatic inflammation. *Int J Biochem Cell Biol* **37**:237–238, 2005.

Leung PS. The physiology of a local renin-angiotensin system in the pancreas. *J Physiol* **580**: 31–37, 2007.

Leung PS and Chan YC. Role of oxidative stress in pancreatic inflammation. *Antioxid Redox Signal* **11**:135–165, 2009.

Lever AF, Hole DJ, Gillis CR, McCallum IR, McInnes GT, MacKinnon PL, Meredith PA, Murray LS, Reid JL and Robertson JW. Do inhibitors of angiotensin-I-converting enzyme protect against risk of cancer? *Lancet* **352**:179–184, 1998.

Li JM and Shah AM. Mechanism of endothelial cell NADPH oxidase activation by angiotensin II. Role of the p47phox subunit. *J Biol Chem* **278**:12094–12100, 2003.

Liu AM and Wong YH. Activation of nuclear factor-kappa B by somatostatin type 2 receptor in pancreatic acinar AR42J cells involves G-alpha 14 and multiple signaling components: a mechanism requiring protein kinase C, calmodulin-dependent kinase II, ERK, and c-Src. *J Biol Chem* **280**:34617–34625, 2005.

Mizuiri S, Hemmi H, Arita M, Ohashi Y, Tanaka Y, Miyagi M, Sakai K, Ishikawa Y, Shibuya K, Hase H and Aikawa A. Expression of ACE and ACE2 in individuals with diabetic kidney disease and healthy controls. *Am J Kidney Dis* **51**:613–623, 2008.

Murr MM, Yang J, Fier A, Kaylor P, Mastorides S and Norman JG. Pancreatic elastase induces liver injury by activating cytokine production within Kupffer cells via nuclear factor-Kappa B. *J Gastrointest Surg* **6**:474–480, 2002.

Muscella A, Greco S, Elia MG, Storelli C and Marsigliante S. Angiotensin II stimulation of Na+/K+ATPase activity and cell growth by calcium-independent pathway in MCF-7 breast cancer cells. *J Endocrinol* **173**:315–323, 2002.

Ocaranza MP, Godoy I, Jalil JE, Varas M, Collantes P, Pinto M, Roman M, Ramirez C, Copaja M, Diaz-Araya G, Castro P and Lavandero S. Enalapril attenuates downregulation of angiotensin-converting enzyme 2 in the late phase of ventricular dysfunction in myocardial infarcted rat. *Hypertension* **48**:572–578, 2006.

Pupilli C, Lasagni L, Romagnani P, Bellini F, Mannelli M, Misciglia N, Mavilia C, Vellei U, Villari D and Serio M. Angiotensin II stimulates the synthesis and secretion of vascular permeability factor/vascular endothelial growth factor in human mesangial cells. *J Am Soc Nephrol* **10**: 245–255, 1999.

Raizada MK and Ferreira AJ. ACE2: a new target for cardiovascular disease therapeutics. *J Cardiovasc Pharmacol* **50**:112–119, 2007.

Rau B, Bauer A, Wang A, Gansauge F, Weidenbach H, Nevalainen T, Poch B, Beger HG and Nussler AK. Modulation of endogenous nitric oxide synthase in experimental acute pancreatitis: role of anti-ICAM-1 and oxygen free radical scavengers. *Ann Surg* **233**:195–203, 2001.

Rivera E, Arrieta O, Guevara P, Duarte-Rojo A and Sotelo J. AT1 receptor is present in glioma cells: its blockage reduces the growth of rat glioma. *Br J Cancer* **85**:1396–1399, 2001.

Sano M, Fukuda K, Sato T, Kawaguchi H, Suematsu M, Matsuda S, Koyasu S, Matsui H, Yamauchi-Takihara K, Harada M, Saito Y and Ogawa S. ERK and p38 MAPK, but not NF-kappaB, are critically involved in reactive oxygen species-mediated induction of IL-6 by angiotensin II in cardiac fibroblasts. *Circ Res* **89**:661–669, 2001.

Schieffer B, Luchtefeld M, Braun S, Hilfiker A, Hilfiker-Kleiner D and Drexler H. Role of NAD(P)H oxidase in angiotensin II-induced JAK/STAT signaling and cytokine induction. *Circ Res* **87**:1195–1201, 2000.

Schulz HU, Niederau C, Klonowski-Stumpe H, Halangk W, Luthen R and Lippert H. Oxidative stress in acute pancreatitis. *Hepatogastroenterology* **46**:2736–2750, 1999.

Suzuki Y, Ruiz-Ortega M, Lorenzo O, Ruperez M, Esteban V and Egido J. Inflammation and angiotensin II. *Int J Biochem Cell Biol* **35**:881–900, 2003.

Tahmasebi M, Puddefoot JR, Inwang ER and Vinson GP. Tissue renin-angiotensin system in human pancreas. *J Endocrinol* **161**:317–322, 1999.

Takeda H and Kondo S. Differences between squamous cell carcinoma and keratoacanthoma in angiotensin type-1 receptor expression. *Am J Pathol* **158**:1633–1637, 2001.

Tamarat R, Silvestre JS, Durie M and Levy BI. Angiotensin II angiogenic effect in vivo involves vascular endothelial growth factor- and inflammation-related pathways. *Lab Invest* **82**:747–756, 2002.

Tanifuji C, Suzuki Y, Geot WM, Horikoshi S, Sugaya T, Ruiz-Ortega M, Egido J and Tomino Y. Reactive oxygen species-mediated signaling pathways in angiotensin II-induced MCP-1 expression of proximal tubular cells. *Antioxid Redox Signal* **7**:1261–1268, 2005.

Telek G, Regöly-Mérei J, Kovács GC, Simon L, Nagy Z and Hamar J, Jakab F. The first histological demonstration of pancreatic oxidative stress in human acute pancreatitis. *Hepatogastroenterology* **48**:1252–1258, 2001.

Tsang SW, Ip SP, Wong TP, Che CT and Leung PS. Differential effects of saralasin and ramiprilat, the inhibitors of renin-angiotensin system, on cerulein-induced acute pancreatitis. *Regul Pept* **111**:47–53, 2003.

Tsang SW, Ip SP and Leung PS. Prophylactic and therapeutic treatments with AT_1 and AT_2 receptor antagonists and their effects on changes in the severity of pancreatitis. *Int J Biochem Cell Biol* **36**:330–339, 2004a.

Tsang SW, Cheng CH and Leung PS. The role of the pancreatic renin-angiotensin system in acinar digestive enzyme secretion and in acute pancreatitis. *Regul Pept* **119**:213–219, 2004b.

Uemura H, Ishiguro H, Nakaigawa N, Nagashima Y, Miyoshi Y, Fujinami K, Sakaguchi A and Kubota Y. Angiotensin II receptor blocker shows antiproliferative activity in prostate cancer cells: a possibility of tyrosine kinase inhibitor of growth factor. *Mol Cancer Ther* **2**:1139–1147, 2003.

Viedt C, Soto U, Krieger-Brauer HI, Fei J, Elsing C, Kübler W and Kreuzer J. Differential activation of mitogen-activated protein kinases in smooth muscle cells by angiotensin II: involvement of p22phox and reactive oxygen species. *Arterioscler Thromb Vasc Biol* **20**:940–948, 2000.

Wang R, Alam G, Zagariya A, Gidea C, Pinillos H, Lalude O, Choudhary G, Oezatalay D and Uhal BD. Apoptosis of lung epithelial cells in response to TNF-alpha requires angiotensin II generation de novo. *J Cell Physiol* **18**:253–259, 2000.

Wang G, Lai FM, Lai KB, Chow KM, Kwan CH, Li KT and Szeto CC. Urinary mRNA expression of ACE and ACE2 in human type 2 diabetic nephropathy. *Diabetologia* **51**:1062–1067, 2008.

Williams B, Baker AQ, Gallacher B and Lodwick D. Angiotensin II increases vascular permeability factor gene expression by human vascular smooth muscle cells. *Hypertension* **25**:913–917, 1995.

Yamanaka K, Saluja AK, Brown GE, Yamaguchi Y, Hofbauer B and Steer ML. Protective effects of prostaglandin E1 on acute lung injury of caerulein-induced acute pancreatitis in rats. *Am J Physiol* **272**:G23–G30, 1997.

Yoshiji H, Kuriyama S and Fukui H. Angiotensin-I-converting enzyme inhibitors may be an alternative anti-angiogenic strategy in the treatment of liver fibrosis and hepatocellular carcinoma. Possible role of vascular endothelial growth factor. *Tumour Biol* **23**:348–356, 2002.

Zhang X, Galardi E, Duquette M, Lawler J and Parangi S. Antiangiogenic treatment with three thrombospondin-1 type 1 repeats vs. gemcitabine in an orthotopic human pancreatic cancer model. *Clin Cancer Res* **11**:5622–5630, 2005.

Zhou L, Zhang R, Yao W, Wang J, Qian A, Qiao M, Zhang Y and Yuan Y. Decreased expression of angiotensin-converting enzyme 2 in pancreatic ductal adenocarcinoma is associated with tumor progression. *Tohoku J Exp Med* **217**:123–131, 2009.

Epilogue

In recent years, there has been emerging evidence of the functional importance of a local renin-angiotensin system in the pancreas and its islets of Langerhans. The local generation of components of this system in the pancreatic islets was only recently described, but seems to affect both vascular and endocrine function in the islets. In addition, the local renin-angiotensin system appears to be critical during the development of disease in pancreatic islets, such as in type 2 diabetes. Moreover, it may restrict the function of transplanted pancreatic islets. The current book provides an excellent overview, and is written by a renowned expert in the area, Po Sing Leung. Since a lot of the research efforts in this field have been conducted by the author himself, the readers should have the opportunity to expose themselves to in-depth knowledge of the field. Further interest and research in this important field are certainly stimulated among readers.

Uppsala, Sweden
December 2009

Per-Ola Carlsson, M.D., Ph.D.

P.S. Leung, *The Renin-Angiotensin System: Current Research Progress in The Pancreas*, Advances in Experimental Medicine and Biology 690,
DOI 10.1007/978-90-481-9060-7,

Epilogue

This book provides an up-to-date overview on the pancreatic renin-angiotensin system (RAS). With the recent introduction of renin inhibitors as the third drug class of drugs to block the RAS, interest in both renin and prorenin has been revived. Prorenin in particular now receives a lot of attention, given its high levels in diabetic subjects experiencing the microvascular complications of this disease. Exactly why prorenin is a marker of these complications, and what its origin in diabetics is, is currently unknown. However, these observations clearly underscore the link between the RAS and diabetes, explaining, at least in part, why RAS blockers are successfully applied in both diabetic nephropathy and retinopathy, and why they affect insulin sensitivity. Knowledge on the pancreas, and the new concept that this organ, like so many others, is capable of generating its own angiotensin II, will allow us to further understand the relationship between diabetes and the RAS. Such knowledge may eventually help to determine precisely what patient will benefit most from which type RAS blocker. This book is an ideal basis to stimulate future research in that direction.

Rotterdam, Netherlands
December 2009

A.H. Jan Danser, Ph.D.

Index

Note: The letter 'f' and 't' followed by the locators refers to figures and tables cited in the text

Breinigsville, PA USA
16 September 2010

245520BV00006B/16/P